THE ESSENCE OF
COMMUNICATIONS THEORY

THE ESSENCE OF ENGINEERING SERIES

Published Titles

THE ESSENCE OF

COMMUNICATIONS THEORY

Richard Read

Prentice Hall Europe

LONDON   NEW YORK   TORONTO   SYDNEY   TOKYO
SINGAPORE   MADRID   MEXICO CITY   MUNICH   PARIS

First published 1998 by
Prentice Hall Europe
Campus 400, Maylands Avenue
Hemel Hempstead
Hertfordshire, HP2 7EZ
A division of
Simon & Schuster International Group

For permission within the United States of America contact
Prentice Hall Inc., Englewood Cliffs, NJ 07632

Typeset in 10/12 pt. Times
by Aarontype Limited, Bristol

Printed and bound in Great Britain by
TJ International Ltd, Padstow, Cornwall

Library of Congress Cataloging-in-Publication Data

Available from the publisher

British Library Cataloguing in Publication Data

A catalogue record for this book is available from
the British Library
ISBN: 0-13-521022-4

1 2 3 4 5   02 01 00 99 98

Contents

Preface

This text is written as an introduction for undergraduates studying communications theory. It offers sufficient detail of the main elements of signals, noise, modulation and the increasingly important topic of digital transmission to enable further study of communication systems. The use of mathematics is dealt with in a friendly style and clearly explained where necessary.

Each chapter commences with aims and objectives and concludes with a summary and end of chapter exercises. In addition each chapter includes worked examples and self-assessment problems which enable the reader to test understanding.

Chapter 1 introduces the absorbing subject of communications. It defines 'communication' and illustrates various facets of the subject through a brief history. A model of a communication system acts as a vehicle to define some basic terminology as well as key elements of a system. The nature of signals is examined and the important distinction made between analogue and digital signals.

Chapter 2 develops signals further and considers common forms of signal processing employed in transmission and reception. The relationship between a signal's representation in time and its frequency content, or spectra, is examined in some detail. This leads naturally to the use of Fourier series to determine the spectrum of a repetitive signal in time. Similarly the Fourier transform is demonstrated to determine the spectrum of a non-repetitive signal. Conversely the inverse Fourier transform shows how a signal's time representation may be found by knowledge of its spectrum. The more advanced topic of convolution is explained clearly and in detail as an alternative, and simpler, technique to multiplying two signals. Analogue filters are introduced to illustrate one of the most basic signal processing techniques. The chapter concludes with an introduction to the important topic of digital signal processing. The key topics of sampling and digital filtering are dealt with.

Chapter 3 provides a standard treatment of noise in communication systems. The chapter commences with a broad examination of the main causes and types of noise. Noise introduced by electronic components is covered in some detail. The concept of effective noise temperature is explained as a means of quantifying the noise contribution a sub-system may make to a system. Noise peculiar to radio systems is examined. Signal to noise ratio, the principal measure of the effectiveness of communication, is defined and numerical examples displayed. Finally, an alternative to effective noise temperature, namely noise figure, is discussed.

Chapter 4 deals with modulation, a process found in many communication systems. Modulation is the process of impressing an information signal upon another waveform, or carrier, and is typically used to enable radio communication. The chapter confines itself to analogue modulation, rather than the use of digital signals which are dealt with in the following chapter. Both amplitude and angle modulation are discussed. Angle modulation may be in the form of phase or frequency modulation. The chapter concentrates on the latter which is far more common. The concept of segregating channels in a multi-channel environment by frequency, known as frequency division multiplexing, is shown. The chapter concludes with a brief comparison of modulation methods regarding both noise and the frequency bandwidth required.

Chapter 5 deals exclusively with digital communication. Increasingly, with the explosive growth of computer communications, signals themselves are in digital form which is the principal reason why digital transmission systems are more important now rather than the analogue systems that were dominant until the 1970s. The chapter commences with the minimum bandwidth necessary to convey a baseband signal. The effect of noise upon such signals is then examined and its effect upon the probability of a bit being received in error. Shannon's channel capacity theorem is stated which indicates the maximum rate at which bits may be transmitted per second over a channel of given bandwidth and signal to noise ratio. Time division multiplexing, an alternative to frequency division multiplexing, is defined as a means of separating channels in time. Pulse code modulation shows how a number of channels may share a channel, making use of time division multiplexing. Each channel periodically transmits a series of bits called a pulse code which represents a sample of the baseband signal. The concept of quantisation error is explained and leads to the development of signal to quantisation noise as a form of 'signal to noise ratio' in such systems. Common forms of digital modulation used in modems, namely amplitude shift keying, frequency shift keying, phase shift keying and quadrature amplitude modulation are discussed. Finally, these modulation techniques are compared.

There is an appendix containing selected answers to both self-assessments and end of chapter exercises. In addition appendices contain Fourier series, Fourier transforms, Bessel functions and the normal error function.

Acknowledgements

The author wishes to acknowledge the numerous helpful and constructive discussions with Mike Duck, Nortel PLC, and Tony Drake, University of North London. Appreciation is also extended to Pat Prigmore for his contribution towards wordprocessing much of the manuscript in its early stages. Thanks are also extended to The Acquisitions Editor Christopher Glennie who originally commissioned this work on behalf of Prentice Hall and also the numerous early draft reviewers, all of whom have have made a most valuable contribution in shaping the final text.

Introduction

Aims and objectives

The term communication is defined and the nature of information to be communicated discussed. A brief history of communications is presented to illustrate the various forms of communication available and some of their basic characteristics. The method of classifying radio waves in terms of frequency and wavelength is indicated, as well as associated ranges of operation and types of services employed. A communication model is presented to introduce the main elements of a communication system. Signals may be classified as continuous or discrete, analogue or digital, and their differences are explained.

1.1 Communication: a definition

Communication may be defined as the process of successfully transferring information between two, or more, parties. Communication may be in one direction only, or bidirectional. The parties themselves may be human, machine or a combination of both.

What is the nature of the information required to be communicated? Consider, for instance, a communications system with which we are very familiar, namely television. Television has similarity with cinematography, or film, in that a succession of images or frames are transmitted, typically at a rate of 25 per second. Fundamentally the information to be transmitted about each frame is intensity of light and colour with its associated position within the frame (or screen). In addition sound must be transmitted. So, in this example, we see that there are three types of information to be transmitted in a television system: light, position and sound.

Communication systems handle information in the form of a **signal**. Modern communication is overwhelming in its use of electronic signals. However light, sound, positional information and many other commonly found sources of information to be communicated do not originate in electrical form. A **transducer** is therefore necessary to convert an information signal to, and from, electrical form. In sound systems transducers are found in the form of microphones and loudspeakers. Television receivers use a cathode ray tube for reproducing a colour image.

1.2 Key historical events

We have touched upon two types of signals commonly transmitted but what of other types of signals? Table 1.1 indicates key landmarks in the historical development of communications. It also serves to introduce a number of other types of signals and systems. Early telegraph systems consisted of manual transmission of two voltage states, for example a positive and a negative voltage. Combinations of these states may be used to create a number of unique **codewords**, each representing a piece of information, typically letters of the alphabet and numbers. An example of such a code is the Morse code. Early telegraph systems could send signals over appreciable distances, many miles, but were characterised by slow speed of operation. The range of frequencies necessary to transmit telegraphy at such speeds is of the order of tens to hundreds of hertz.

Telegraphy communication systems remain in existence today. They are mainly found in applications such as maritime communications using radio communication rather than the original wire-based systems of the last century. They may also be found in Telex systems which are a dial-up type of network, rather like a telephone system. Earlier Telex systems used teleprinters, enabling a message typed on one machine to be printed on paper at a distant machine. More usually today, Telex is run as an application upon a PC (personal computer) platform. Telex systems are still of value in certain commercial applications where modern computer communications, such as Internet, are not readily available. Typical speed of operation today is of the order of five alpha-numeric characters per second, using automated transmission in contrast to the slower manual keysender.

The next major advance was the introduction of the telephone by Bell in 1876. Early telephone systems operated over lines rather than radio, which was

Table 1.1 Key developments in communications

Year	Event
1838	Introduction of the telegraph (Samuel Morse)
1876	Invention of the telephone (Alexander Graham Bell)
1897	Invention of wireless telegraph (Guglielmo Marconi)
1906	Triode vacuum tube (valve) Lee De Forest
1938	Television broadcasting commences at Alexander Palace, London
1940–45	Radar developed
1948	Transistor invented
1959	Integrated circuit (IC) invented
1962	Telstar, communication satellite, launched. Heralded start of global transmission of live television
1968	Optical fibre communication introduced

still in its infancy. The frequency range of human speech, and therefore the electrical signals used to convey them, lies in the range 100 Hz to 10 kHz, or so, which is an order greater than that of telegraph signals. The telephone is limited by the fact that electrical signals become weaker, or **attenuate**, with distance. Later, in 1906, electronic amplification of signals using valves (and transistors after 1948) opened up the possibility of long-distance communication over, for example trans-Atlantic submarine cables.

Marconi's experiments in radio in the late nineteenth century gave way initially to transmission of telegraph signals, and later speech, over distances far in excess of what was then possible using conducting wires. Indeed radio signals became routinely transmitted across the Atlantic. At certain frequencies truly global transmission is possible.

Table 1.2 illustrates how various bands of electromagnetic, or radio, waves are classified, and their corresponding range of frequency and wavelength. In addition the table shows the range of operation of each band and some of the uses that are made of them.

Wavelength and frequency are directly related, as the following Example illustrates.

EXAMPLE 1.1
Determine the wavelength of a radio wave of frequency 20 MHz

SOLUTION
The velocity of propagation of an electromagnetic wave in free space is 3×10^8 m/s and is usually denoted by the symbol c. (Note, it is usual to assume a similar figure in the earth's atmosphere although in practice velocity is slightly less.)

$$c = \frac{f}{\lambda} \tag{1.1}$$

$$\therefore \quad \lambda = \frac{f}{c} \tag{1.2}$$

Hence

$$\lambda = \frac{3 \times 10^8}{20 \times 10^6} \tag{1.3}$$

$$= 15 \,\text{m} \tag{1.4}$$

Self-assessment 1.1 A certain radio wave has a wavelength of 40 cm. Calculate its frequency.

After the introduction of the telegraph, the telephone and radio, the last two later making extensive use of amplification, the next major milestone was that

Table 1.2 Classification of radio waves

Band designation	Frequency range	Wavelength	Range	Services
VLF (very low frequency)	3–30 kHz	100–10 km	World-wide	Long distance communication, navigation, slow-speed data
LF (low frequency)	30–300 kHz	10–1 km	World-wide	As for VLF
MF (medium frequency)	300–3000 kHz	1–0.1 km	Hundreds of kilometres	Entertainment broadcasting (audio), maritime communication
HF (high frequency)	3–30 MHz	100–10 m	World-wide by multiple hops	Medium and long-distance point to point and broadcast
VHF (very high frequency)	30–300 MHz	10–1 m	Line of sight	Short-range point to point and broadcast (paging, audio and TV); medium-range point to point by tropospheric scattering; mobile radiocommunication
UHF (ultra high frequency)	300–3000 MHz	1–0.1 m	Line of sight	As for VHF; satellite communication
SHF (superhigh frequency)	3–30 GHz	10–1 cm	Line of sight	Short-range terrestrial point to point; satellite – point to point, mobile and broadcast; radar
EHF (extremely high frequency)	30–300 GHz	1–0.1 cm	Line of sight	Point to point terrestrial links

of television transmission in 1938. A television signal produced by a camera contains frequency components as high as several MHz. Television transmissions are, in principle, similar to those of any other signal that makes use of radio in that the signal is superimposed onto a radio frequency carrier. As is explained in Chapter 4, for proper operation, the carrier frequency must be many times greater than that of the signal to be transmitted. What makes television broadcasting such a step forward is that because of the high frequencies of television signals (which are two orders greater than those of earlier communication systems), a radio frequency carrier in the VHF (very high frequency) band, or higher, must be chosen. This means that the frequency range that the associated radio equipment must be capable of handling is far greater than anything else previously seen.

During World War II the high frequency advances made in the development of television were further exploited for Radio Detection And Ranging (RADAR) purposes, especially for use with aircraft. RADAR, although using radio waves at VHF and UHF (ultra high frequency), also heralded the use of radio waves at even higher **microwave** frequencies. Such frequencies are generally regarded to be those in excess of about 500 MHz where the wavelengths become relatively short, tens of cm or less, hence the use of the term.

By the 1950s long-distance communication of speech was possible. Trans-Atlantic telephone calls were commonplace. Transmission of speech signals may be made by either long-haul submarine cable where available, or by means of radio links. Communication of television over cable, because of its high frequency signal, attenuated rapidly with distance, precluding long-distance communication. Television transmission at VHF and UHF is, as indicated in Table 1.1, governed by line of sight operation. Even if relatively tall radio masts are used, distance is limited to about 50 km.

In the 1960s, with the launch of Telstar and subsequent communications satellites, long-distance television transmission became a reality. This was made possible by using integrated circuits (ICs) to reduce the physical size and weight of electronic systems as well as recent advances in rocket technology enabling larger payloads to be launched. Such satellites are capable of handling the large frequency range of a television signal and relaying signals between continents. They may also provide a near global coverage.

Optical fibre transmission medium, as an alternative to metallic conductors generally of copper, are able to handle all the signals commonly found today including those of television. In consequence there is now a healthy competition between satellite and submarine optical fibre cables for long-distance communication provision.

In the 1970s traditional telecommunications became pervaded by powerful VLSI (very large-scale integration) ICs performing increasingly complex processing operations. Telephone exchanges, for instance, became very large computer-controlled switches. So began the convergence between the telecommunications and computing disciplines. In parallel with this, computers became increasingly powerful and computer inter-networking was introduced on a large scale.

The developments of the 1970s were built on in the 1980s. Mobile telephone networks, based upon cellular radio principles, became commonplace. These networks were made possible because a large degree of intelligence was available for the complex tasks to be performed within handsets, at the radio base station controllers and of course the large switching infrastructure required.

The 1990s have seen further development, arising from mobile communication, of personal communication. When fully developed, personal communication networks (PCNs) will provide an individual with all of the various communication facilities wherever he or she is geographically positioned at any particular instant. That is, all of the personal communications needs will be provided, on a truly mobile basis, around the globe using a universal access technique and numbering system.

1.3 A communication model

This section establishes a model of a communication system. Such a model is useful both to introduce the main elements of a communication system and to define general terminology which will be used throughout the following chapters.

Figure 1.1 illustrates a simple model of a communication system. In many systems the source (and destination) information is non-electrical in form. As we saw earlier, information such as sound is a form of mechanical vibrations. Television is primarily concerned with light energy. A **source encoder**, or **transducer**, is often required to convert the source information into electrical form. The electrical form of the information, or message, is known as the **message signal** and which we shall denote throughout this text by the function of time $m(t)$. Similarly, at a receiver, a source decoder, or transducer, is again required to convert the message signal from an electrical form into some other from of energy to reproduce the original information. Two common transducers are a television camera and a loudspeaker. The channel's response in time is represented by the function $h(t)$ and $k(t)$ represents the output signal from the channel.

A **transmitter** is used to convert the message signal into a form acceptable to the channel and which we shall represent as $g(t)$; a **channel** is the path, or link, that interconnects transmitter and receiver. As we have already seen the channel may be metallic, optical or radio. A **receiver** performs the converse function to that of the transmitter to recover the message signal.

The model is ideal in the sense that each element is perfect and does not introduce any extraneous energy, which is regarded as **noise**, a subject dealt with in detail in Chapter 3. It is also assumed that no **interference** is introduced from other sources.

1.4 Classification of signals

Information represented in communication systems, in electrical form, may either vary in a **continuous** manner or only change at **discrete** intervals in time. Examples of continuous time signals are speech, audio and video, whereas

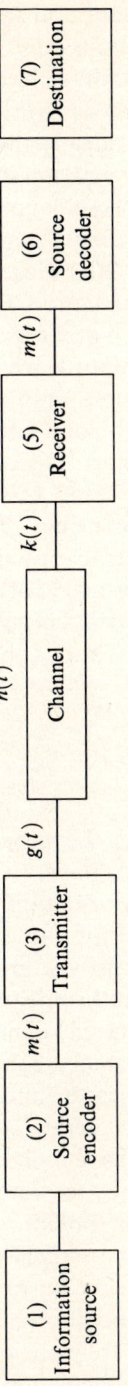

Figure 1.1 Communication system

discrete time signals are found in data and sampled systems such as digitised speech and image systems. Continuous time signals, within a certain range, have an infinite number of possible voltage levels. Speech, for instance, appears as air pressure variation and a typical waveform and equivalent electrical signal are shown in Fig. 1.2(a). Note that ideally the electrical signal varies in exact sympathy with that of the pressure variation due to speech. In other words the electrical signal is an **analogue** of, or analogous to, that of speech.

A simple example of a discrete time signal is the output of an analogue–digital convertor (ADC). Figure 1.2(b) illustrates how, as the first stage of digitising a signal, for example in an audio CD (compact disc) system, a series of voltage samples are obtained, each at a regular and discrete interval in time.

Signals that are not analogue in nature are known as **digital signals** and are predominantly binary in nature where only two possible voltages exist, e.g. binary 0 is represented by 0 V and binary 1 by +5 V. Such a two-state signal is called a **binary digit**, or **bit**, for short. Digital signals clearly do not have an infinite number of possible values or states as with analogue signals. In order to represent a wide range of information using only two states, states must be used in combination. (An alternative sometimes used is that of multi-state signals, for example consisting of four voltage levels.) More commonly short groups of bits, such as a byte, are formed, each representing a unique piece of the original information, e.g. by using eight bits it is possible to represent 2^8, or 256, states.

1.5 Summary

Communication may be defined as the process of successfully transferring information between two, or more, parties. Information from a variety of sources is communicated. Where information occurs in non-electrical form a transducer is employed to convert it into an electronic signal.

A brief history of communications was presented to illustrate the various forms of communication available and some of their basic characteristics. Key developments are radio to support wireless and long-distance communication. Amplifiers, also to support long-distance communication, are used for both cable and radio to overcome loss with distance.

Radio waves are classified as a series of frequency, or corresponding wavelength, bands. Different bands have different properties in regard to range of operation and the types of service to which they may be applied. Below VHF, long-distance, and on occasion global, radio communication is possible. However, the correspondingly lower frequency carriers preclude larger bandwidth signals such as television being carried. VHF, and beyond, is predominantly limited to line of sight operation but can support wideband signals such as television. At microwave frequencies long-haul communication is available by means of satellites. Satellite communication was made possible by

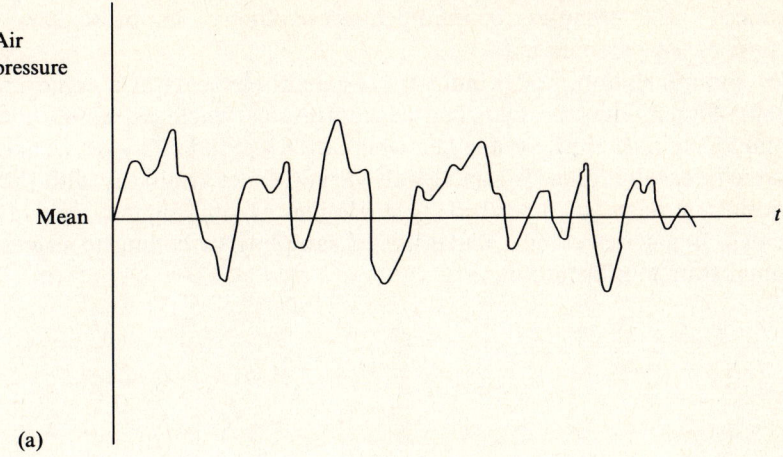

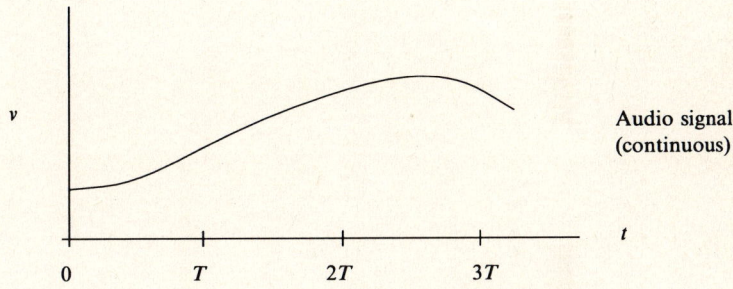

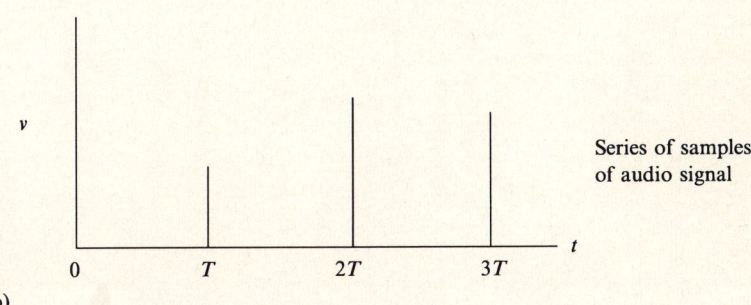

Figure 1.2 (a) Continuous and (b) discrete time signals

the advent of the IC. This invention has subsequently led to the development of advanced and complex communication systems; in particular, mobile telephones for personal use.

A communication model indicates the main elements of a communication system. Signals may be classified as continuous where an amplitude varies continuously over time, or discrete where an amplitude may only change at a discrete interval in time. Where signals vary in exact sympathy with the source information they are described as analogue. Digital signals, generally represented as a series of bits, make use of sampling or coding to represent the original source information.

CHAPTER 2

Signal processing

Aims and objectives

*In order that signals are communicated successfully locally, around the globe or between earth and systems located in space, many different processing operations are necessary. **Signal processing** is concerned with the presentation of signals, in a suitable form, for onward transmission through one or more systems. Often signals need to be processed into other forms, or representations, in order to be finally retrieved at a distant location in, or as nearly as possible, their original form.*

The concept of frequency spectrum is introduced. The spectrum of a signal has profound implications which must be considered in any signal communication system. This chapter shows that a signal's time and frequency domain representations are interrelated. The use of Fourier techniques is introduced for determination of the frequency spectrum of a signal by knowledge of its time domain representation, and vice versa for periodic and non-periodic signals. A number of standard signal types are introduced and described in pictorial form, as well as some basic signal processing operations. The concept of filtering, a standard signal processing operation found in most communication systems, is illustrated. The key operation of convolution, which is essential for communications engineers to master, is described and its practical relevance explained.

*Digital signal processing is introduced whereby signals are **digitised**, enabling signals to be processed by means of computer and microprocessor type devices using numerical methods. The same operations as analogue processing are possible digitally. The main digital signal processing technique demonstrated is that of digital filtering.*

2.1 Frequency spectra

Signals found in communication systems are complex waveforms. In many instances systems may be analysed using a single sine wave input. We shall, in general, let $g(t)$ represent a signal. If $g(t)$ is a sine wave, then we may represent this as:

$$g(t) = A\sin(\omega t + \phi) \qquad (2.1)$$

where A is the sinusoid's amplitude,
        ω is the angular velocity of the sinusoid in radian/s, and
        ϕ is an arbitrary phase in radian.

Note that $\omega/2\pi$ is the frequency of the sinusoid in hertz. The expression for $g(t)$ shown in Eqn (2.1) may be shown pictorially as shown in Figure 2.1, either graphically or as a phasor, and where ϕ is the phase of the sinusoid at time t equal to 0.

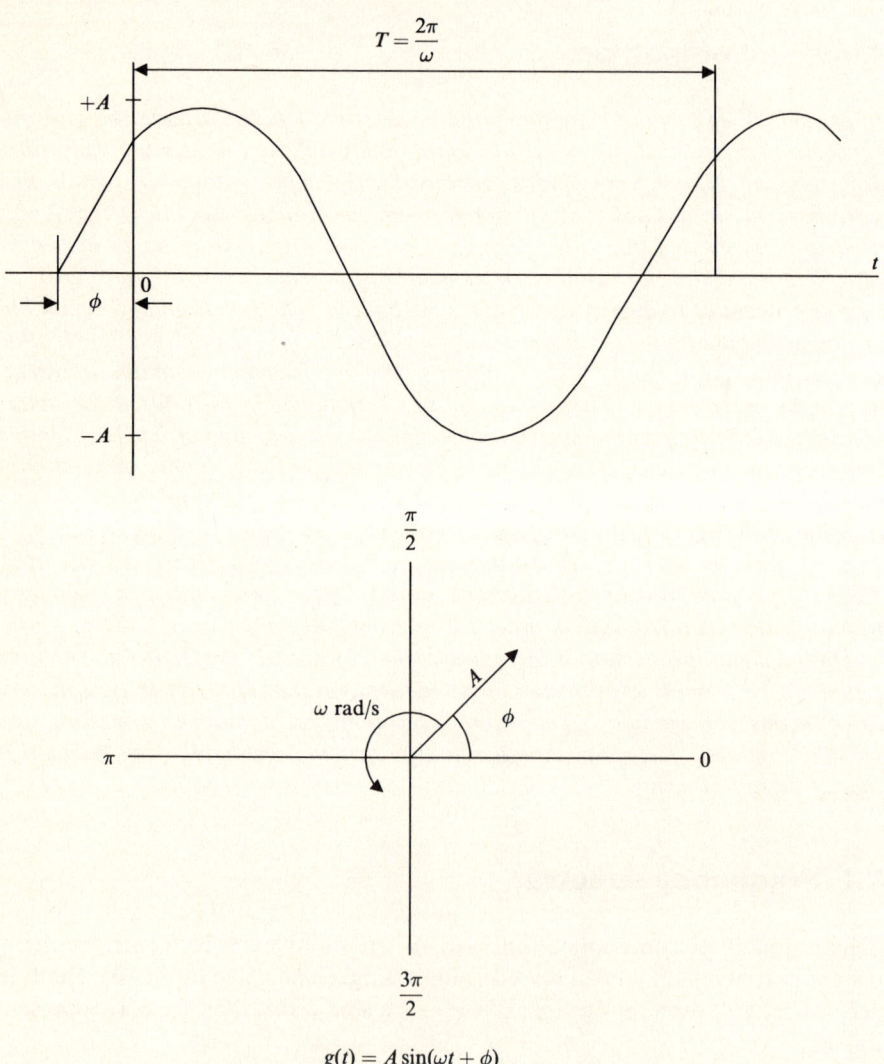

$$g(t) = A\sin(\omega t + \phi)$$

Figure 2.1   Pictorial representation of $g(t)$

We have already mentioned that a transmitter is required to convert signals into a suitable form for transmission over a channel. A key operation is to change the signal's spectral properties, or frequency range, to match the range of frequency, or **bandwidth**, offered by the channel. In some systems the message signal's frequency content or **frequency spectrum** may be applied directly to a channel. This is known as **baseband** operation. In the case of a radio or optical fibre link, message signal frequency components are usually different from those that the channel is able to carry. In such cases the baseband frequencies must be translated to much higher frequencies. Frequency translation, by a process known as **modulation**, is discussed in detail in Chapter 4 which deals with analogue signals, and Chapter 5 for the case of digital signals.

In order to determine whether a signal's spectrum matches that of a channel we must be able to determine its spectrum. A number of 'standard' signals have a known spectrum. For other signals we must have tools for working them out. Consider for instance the sinusoid that we looked at in Eqn 2.1:

$$g(t) = V\sin(\omega t + \phi) \tag{2.2}$$

In this case we know that $g(t)$ contains only one frequency component, namely $\omega/2\pi$ Hz. As an alternative to the time domain representation shown in Figure 2.1, $g(t)$ may alternatively be represented in the frequency domain, Figure 2.2(a), where it is denoted by the term $G(f)$. The vertical axis represents amplitude and the horizontal axis frequency. We may see that the spectrum $G(f)$ has but one component which is known as an **impulse** or **delta** function. This type of spectrum is also termed a **line spectrum** for obvious reasons and is an example of a **single-sided** spectrum. Figure 2.2(b) shows $G(f)$ in **double-sided** form where the signal energy is shared equally at $-\omega/2\pi$ Hz and $+\omega/2\pi$ Hz.

The spectrum shown in Figure 2.2 is an example of a **discrete** spectrum where energy resides at a single, or discrete, frequency. In this case the time domain waveform that produced a discrete spectrum is **continuous** for all time, both negative (past) and positive (future). In fact any signal that is continuous in time has a discrete spectrum: that is, its spectrum does not span all frequencies. The converse is also true in that a discrete time signal, such as a single pulse, has a spectrum that is continuous, as will be seen later in Section 2.3.

The spectral content of a complex wave that is **periodic**, i.e. a short section of the wave that occurs repeatedly as shown in Figure 2.3, can be represented mathematically as a **Fourier series**. The spectrum of an **aperiodic** waveform, i.e. one that is not repetitive such as a single pulse, may be found by taking the **Fourier transform** of the mathematical expression defining the waveform in time.

(a)

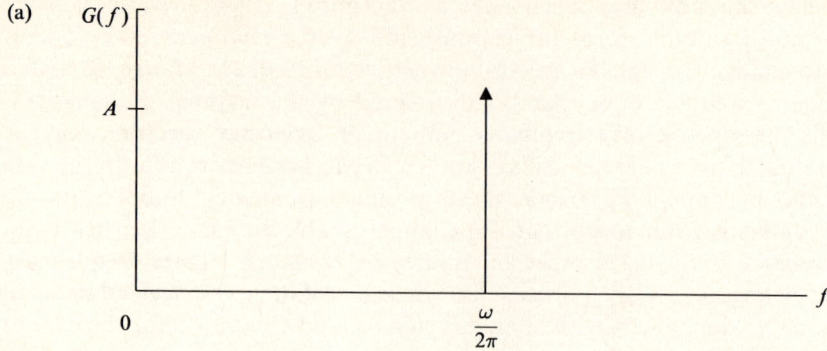

(b)

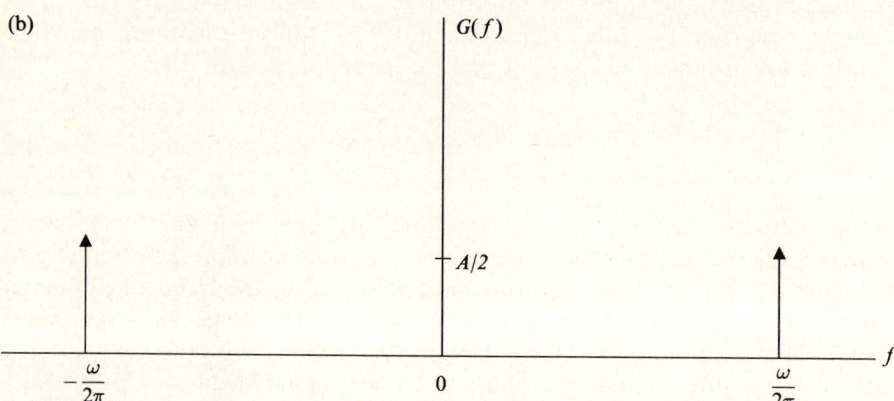

Figure 2.2   Sine wave in the frequency domain, $G(f)$: (a) single sided; (b) double sided

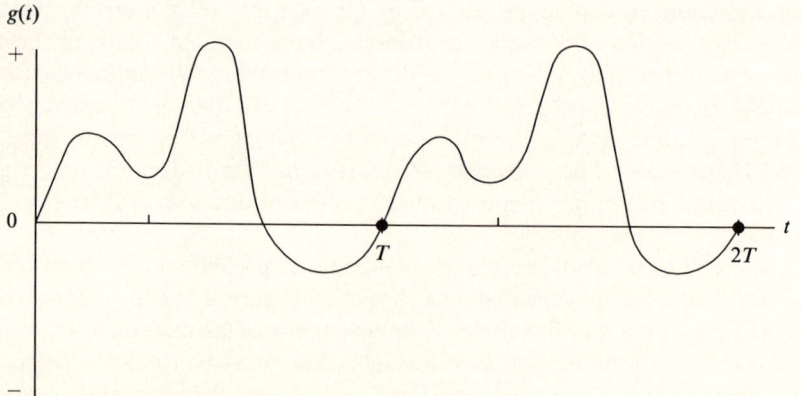

Figure 2.3   A periodic waveform

2.2 Fourier series

The periodic waveform shown in Figure 2.3 has period T and a frequency of repetition f equal to $1/T$. A periodic waveform $g(t)$ with period T equal to $2\pi/\omega$ has, in general, an infinite spectrum expressed by the Fourier series:

$$g(t) = a_0 + a_1 \cos \omega t + b_1 \sin \omega t + a_2 \cos 2\omega t + b_2 \sin 2\omega t + \cdots \qquad (2.3)$$

The first term a_0 is a dc (direct current) term which may, or may not, exist. There are then two components, $a_1 \cos \omega t$ and $b_1 \sin \omega t$, at frequency $\omega/2\pi$, that have the same frequency as that of the complex wave, that is reciprocal of its period of repeat or period. This frequency is said to be the **fundamental** frequency. The next two components are at twice the fundamental frequency and are termed the **second harmonic**. Although not shown, in general, the series continues to infinity with further multiples of the fundamental frequency, i.e. third harmonic, fourth harmonic, etc.

The above series may be more succinctly represented mathematically:

$$g(t) = a_0 + \sum_{n=1}^{\infty} a_n \cos n\omega t + \sum_{n=1}^{\infty} b_n \sin n\omega t \qquad (2.4)$$

The a_n, b_n coefficient pairs with their associated cos and sin terms are merely a quadrature representation of a single rotating phasor, already seen in Figure 2.1, with amplitude c_n and angular velocity ω rad/s. Each c_n has an arbitrary phase when t equals zero which is found thus:

$$\phi_n = \arctan \frac{a_n}{b_n} \qquad (2.5)$$

The amplitude c_n is found:

$$c_n = \sqrt{a_n^2 + b_n^2} \qquad (2.6)$$

We may now represent each sinusoid–cosinusoid pair for a each harmonic by a single rotating phasor shown in Figure 2.4. This now means that we may alternatively rewrite Eqn (2.4) more succinctly:

$$g(t) = c_0 + \sum_{n=1}^{\infty} c_n \sin(n\omega t + \phi_n) \qquad (2.7)$$

What distinguishes the spectrum of one periodic waveform from that of another are the values of the Fourier coefficients a_n and b_n. They are found

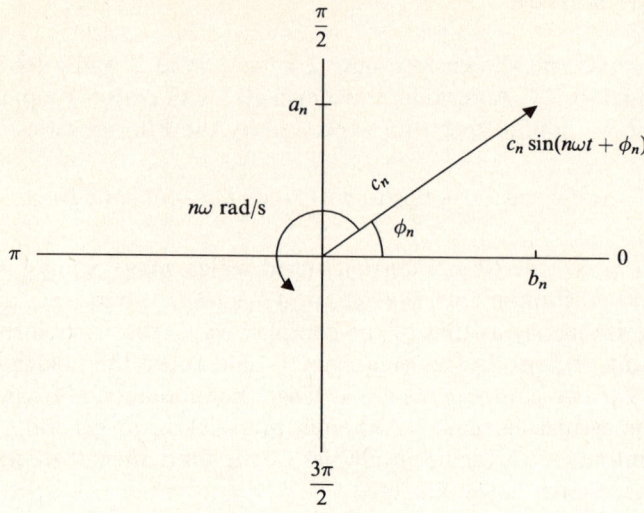

Figure 2.4   A single harmonic

using the following equations:

$$a_0 = c_0 = \frac{1}{T} \int_0^T v(t)\,dt \tag{2.8}$$

$$a_n = \frac{2}{T} \int_0^T v(t)\cos n\omega t\,dt \tag{2.9}$$

$$b_n = \frac{2}{T} \int_0^T v(t)\sin n\omega t\,dt \tag{2.10}$$

Note that a_0 is simply the mean value of one complete cycle.

EXAMPLE 2.1
Determine the spectrum of the waveform shown in Figure 2.5.

SOLUTION
By inspection the mean value, or dc component, is 0 V, i.e. $a_0 = 0$.
Alternatively we could have used Eqn (2.8) to calculate a_0, the dc value.
   Next we must define an expression for $g(t)$ over one complete cycle,
that is duration T:

$$\begin{aligned} g(t) &= +V, & 0 \prec t \preceq T/4 \\ &= -V, & T/4 \prec t \preceq 3T/4 \\ &= +V, & 3T/4 \prec t \preceq T \end{aligned}$$

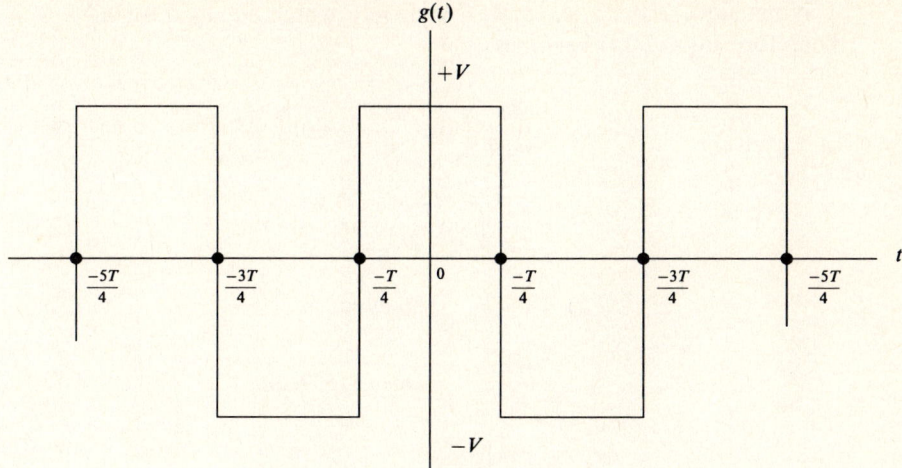

Figure 2.5

Now to determine the cosinusiodal components of the Fourier series:

$$a_n = \frac{2}{T} \int_0^T V(t) \cos n\omega t \, dt \tag{2.11}$$

$$= \frac{2}{T} \int_0^{T/4} V \cos n\omega t \, dt + \frac{2}{T} \int_{T/4}^{3T/4} -V \cos n\omega t \, dt$$

$$+ \frac{2}{T} \int_{3T/4}^T V \cos n\omega t \, dt \tag{2.12}$$

$$= \frac{2V}{T} \left(\frac{1}{n\omega} \sin n\omega t \right)_0^{T/4} + \left(\frac{-1}{n\omega} \sin n\omega t \right)_{T/4}^{3T/4}$$

$$+ \left(\frac{1}{n\omega} \sin n\omega t \right)_{3T/4}^T \tag{2.13}$$

$$= \frac{2V}{n\omega T} \left[\sin \frac{n\omega T}{4} - 0 - \sin \frac{n\omega 3T}{4} + \sin \frac{n\omega T}{4} \right.$$

$$\left. + \sin n\omega T - \sin \frac{n\omega 3T}{4} \right] \tag{2.14}$$

Now, since $T = 2\pi/\omega$, $\sin n\omega T = \sin n2\pi$ which equals 0 for all n. Therefore Eqn (2.14) becomes:

$$a_n = \frac{4V}{n\omega T}\left(\sin\frac{n\omega T}{4} - \sin\frac{n\omega 3T}{4}\right) \tag{2.15}$$

$$= \frac{2V}{n\pi}\left(\sin\frac{n\pi}{2} - \sin\frac{n3\pi}{2}\right) \tag{2.16}$$

$$\therefore \quad a_n = 0, \qquad n \text{ even}$$

$$= \frac{4V}{n\pi}, \qquad n = 1, 5, 9, \text{etc.}$$

$$= \frac{-4V}{n\pi}, \qquad n = 3, 7, 11, \text{etc.}$$

Now we shall determine the sinusoidal components:

$$b_n = \frac{2}{T}\int_0^T V(t)\sin n\omega t\, dt \tag{2.17}$$

$$= -\frac{2V}{n\omega T}[(\cos n\omega t)_0^{T/4} + (\cos n\omega t)_{T/4}^{3T/4} + (\cos n\omega t)_{3T/4}^T] \tag{2.18}$$

$$= \left[\frac{-2V}{n\omega T}\cos\left(\frac{n\omega T}{4}\right) - \cos 0 + \cos\left(\frac{n\omega 3T}{4}\right)\right.$$

$$\left. - \cos\left(\frac{n\omega T}{4}\right) + \cos(n\omega T) - \cos\left(\frac{n\omega 3T}{4}\right)\right] \tag{2.19}$$

$$= \frac{-2V}{n\omega T}(-1 + \cos n\omega T) \tag{2.20}$$

where

$$\omega = \frac{2\pi}{T}$$

$$\therefore \quad b_n = \frac{-2V}{n\omega T}(-1 + \cos n2\pi)$$

$$= 0 \tag{2.21}$$

Hence the spectrum has only cosinusoidal components, there being no sinusoidal components, nor a dc component. Of the sinusoidal components, only **odd harmonics** are present with alternate sign, or polarity, change. Finally, the expression for $g(t)$ may be written:

$$g(t) = \frac{4V}{\pi} \left(\cos \omega t - \tfrac{1}{3} \cos 3\omega t + \tfrac{1}{5} \cos 5\omega t - \cdots \right) \qquad (2.22)$$

The spectrum is as illustrated in Figure 2.6 below.

The calculation of coefficients a_n and b_n in the above example could have been made simpler. Although, in Eqns (2.8) (2.9) and (2.10), the limits were defined as 0 and T, this does not have to be the case. All that is important in using Fourier series is that integrals are taken over exactly one cycle of the waveform. Calculation could have been reduced by using limits $-T/4$ to $+3T/4$, i.e. the waveform $g(t)$ could then be defined by only two expressions, rather than three that we used.

Self-assessment 2.1 Determine the spectrum of the waveform shown in Figure 2.7.

For many simple periodic waveforms it is not necessary to calculate Fourier series coefficients. Rather, as shown in Appendix B, commonly found waveforms have coefficients that are well known and documented.

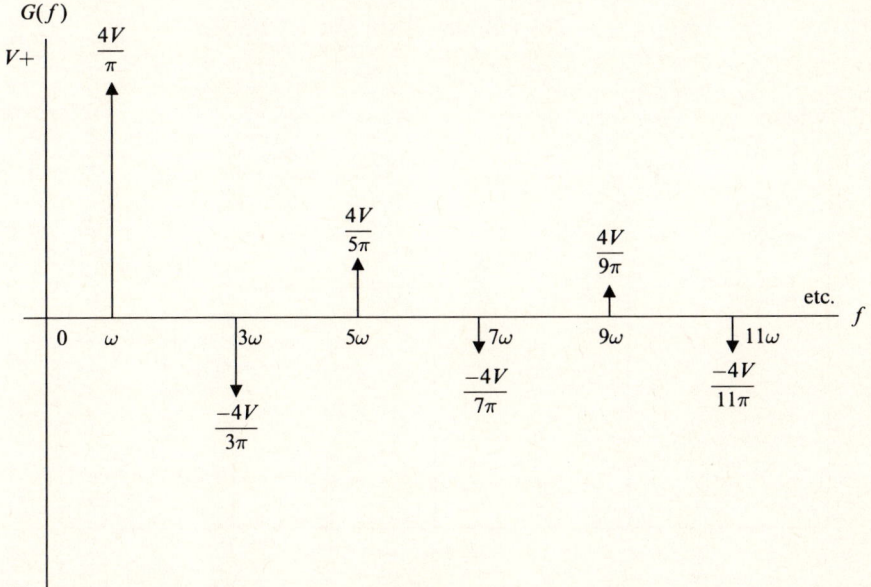

Figure 2.6

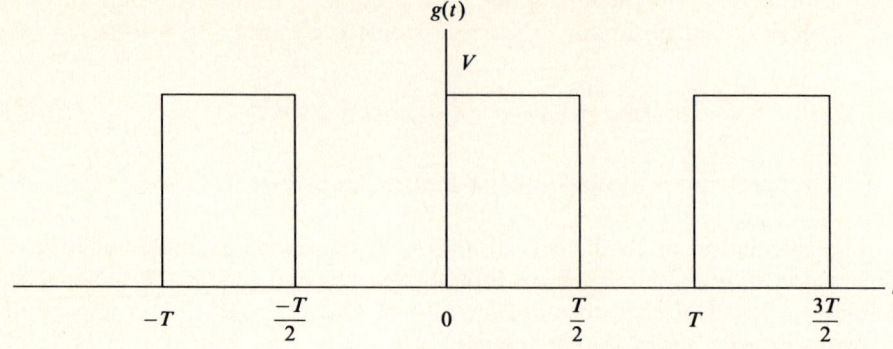

Figure 2.7

EXAMPLE 2.2
Determine the spectrum for the square wave shown in Figure 2.8 by means of Fourier series tables.

SOLUTION
We may see that the waveform has a mark to space ratio of 1:1 and period T. By inspection, the dc term a_0 is in this case $V/2$. If we remove the dc component we are left with a waveform that is identical in form to that shown in B.2 of Appendix B with a peak value of $V/2$. Hence $g(t)$ is:

$$g(t) = \frac{V}{2} + 2\,\frac{V}{\pi}\,(\sin \omega t + \tfrac{1}{3} \sin 3\omega t + \tfrac{1}{5} \sin 5\omega t + \cdots) \qquad (2.23)$$

where $\quad \omega = \dfrac{2\pi}{T}$ $\qquad\qquad\qquad\qquad\qquad\qquad\qquad\qquad\quad$ (2.24)

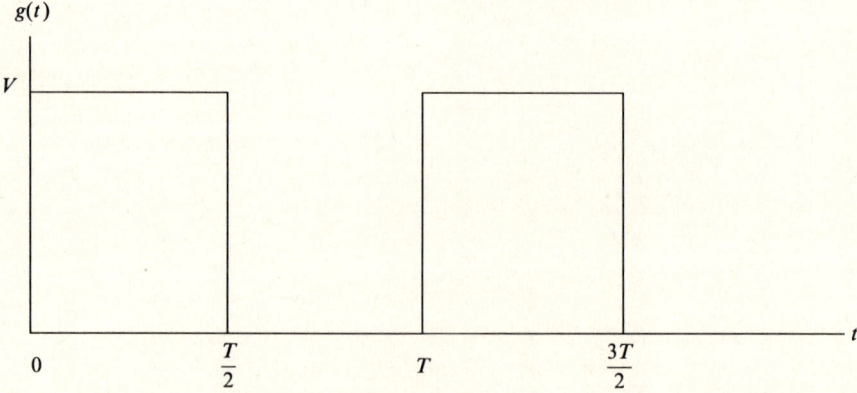

Figure 2.8

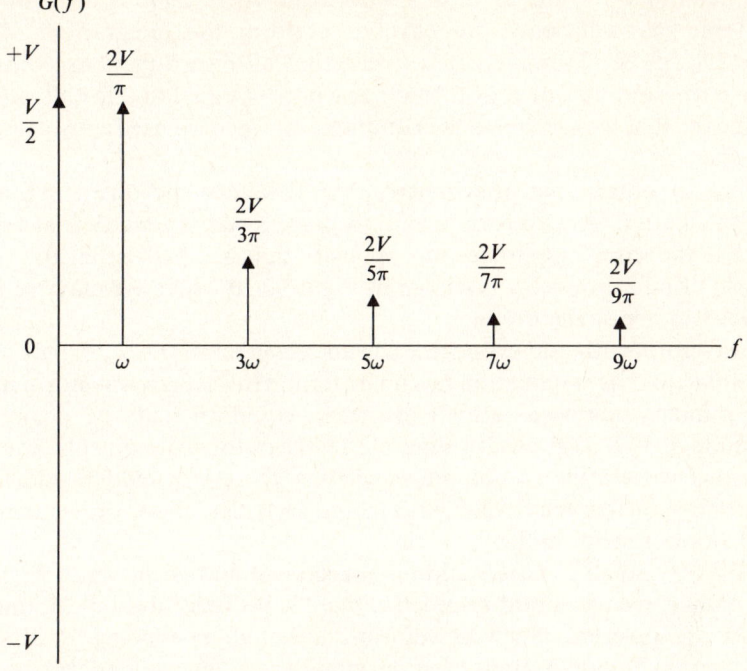

Figure 2.9

The function $g(t)$ contains only odd sinusoidal components and a dc term of $V/2$ and is shown in Figure 2.9.

This example also demonstrates how we may analyse a waveform, using **superposition**, as the sum of two, or more, other waveforms. In this case $g(t)$ is the sum of a dc term of $V/2$ and a bipolar square wave of peak value $V/2$. Their two separate spectra may be found and then added to produce the final spectrum, i.e. a function of amplitude $V/2$ at 0 Hz plus odd sinusoidal harmonics. This is a useful concept to grasp. A waveform to be analysed may not be a standard form, precluding direct use of Fourier series tables. However, it may be that on closer inspection it may be viewed as the sum of a number of waveforms which are standard forms, each of which may be dealt with using tables. The constituent spectra may then, by application of superposition, be simply summed.

The waveform in Example 2.1 earlier, and shown in Figure 2.5, is an example of an **even function**; that is, one that is reflected about the vertical axis. Mathematically we can describe an even function $f(t)$ as being equal to $f(-t)$; that is, the function has the same value at time t as it does at time $-t$. An even function may, or may not, contain a dc term.

To determine the spectrum of a waveform where there is no readily available table of coefficients, we have to perform the integrations shown in Eqns (2.8)–(2.10). However, it is a rule that all even functions contain only cosine terms and no sine terms. In other words b_n equal to 0 for all values of n. This means that we can halve the computation since we need only calculate the a_n coefficients.

There are other types of symmetry that, if we can spot them, can also save some calculation. A waveform is said to possess **odd** symmetry if at time t and $-t$ it has the same amplitude, but opposite sign, i.e. $f(t)$ equal to $-f(t)$. An example of odd symmetry is shown in Figure 2.10 where we may see that it is symmetrical about the origin.

A waveform with odd symmetry has no cosine terms, i.e. a_n equal to 0 for all values of n, and by implication has no dc term. However, a waveform may have odd symmetry imposed upon a dc term, equal to half the peak to peak amplitude or $V/2$, as shown in Figure 2.11. Therefore care must be exercised in examining waveforms to determine whether there is a hidden odd function. Failure to spot this may result in unnecessary calculation, rather than simple use of Fourier series tables.

A third symmetry classification is **pseudo-odd** and is shown in Figure 2.12. Here, unlike even and odd symmetry, there is no reflection but alternate half-periods are inverted. Such waveforms, although possessing both sine and cosine terms, do not contain even harmonics nor hence a dc term.

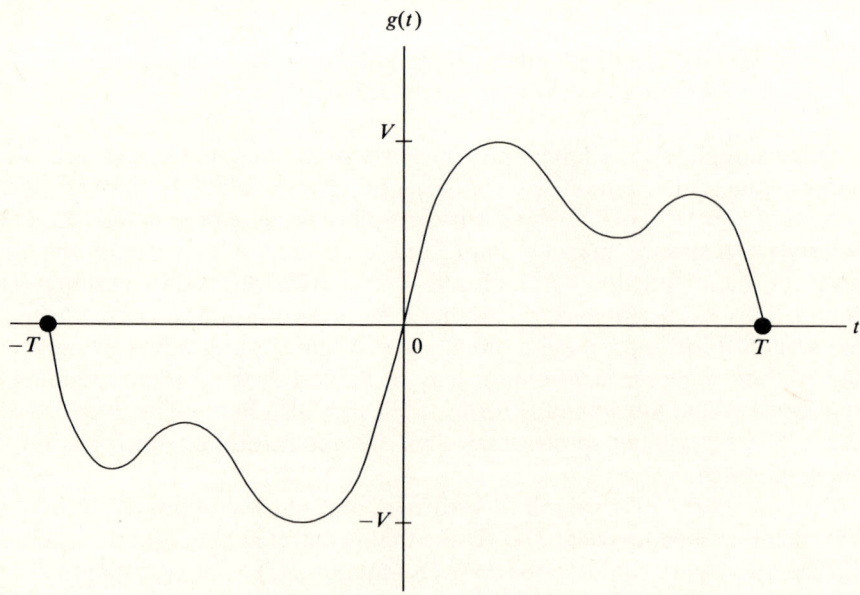

Figure 2.10   Odd symmetry

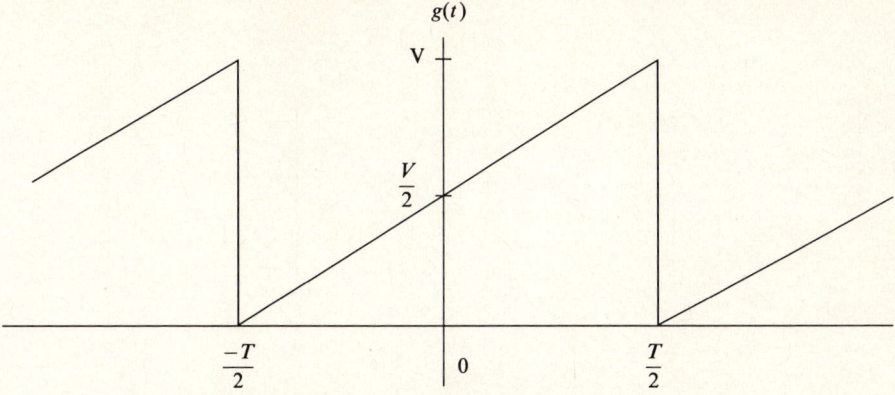

Figure 2.11 Odd symmetry superimposed on a dc term

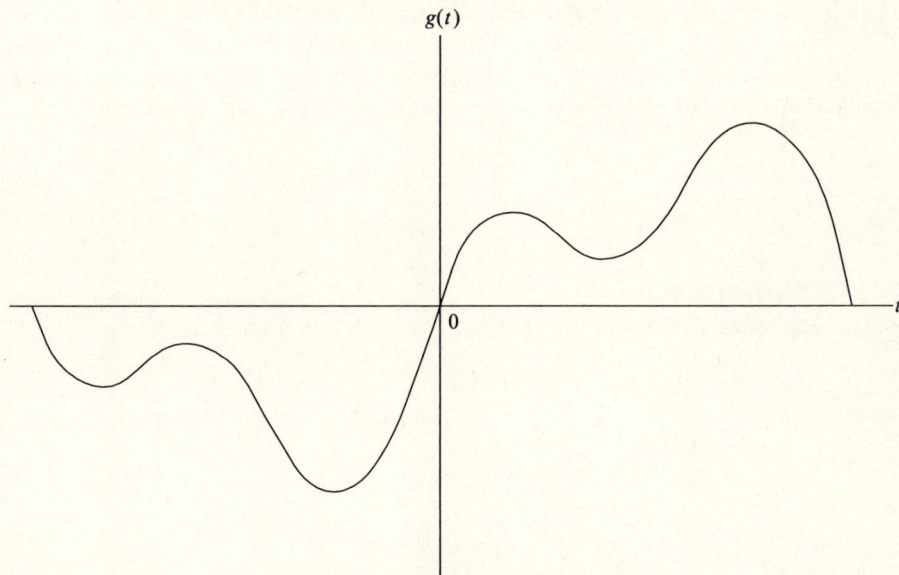

Figure 2.12 Pseudo odd symmetry

Self-assessment 2.2 Determine the spectrum of the waveform shown in Figure 2.13 by use of Appendix B, rather than first principles.

2.3 The Fourier transform

Use of Fourier series enables determination of the spectrum of a periodic waveform which is continuous for all time from $t = -\infty$ to $t = +\infty$. In reality

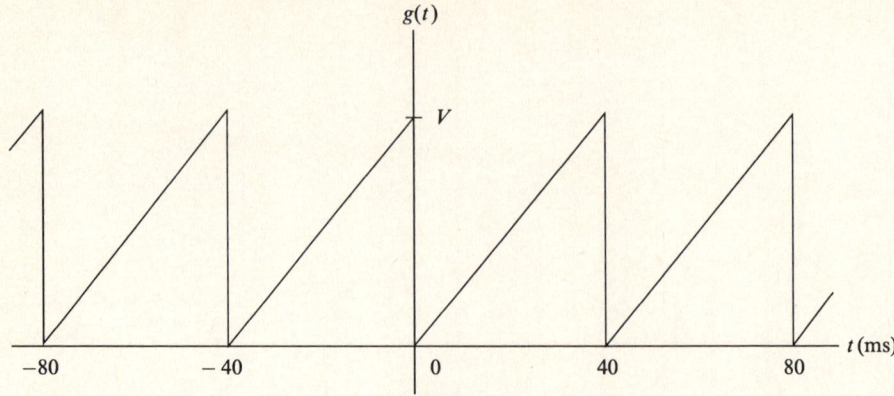

Figure 2.13

signals can never be continuous but if a signal's spectrum is assumed to be so, the error introduced is in most practical cases acceptable. In order to find the spectrum of a signal which is not continuous such as a single pulse, i.e. a time-limited function, the **Fourier transform** may be used which is defined:

$$G(f) = \int_{-\infty}^{\infty} g(t)\,e^{-j2\pi ft}\,dt \qquad (2.25)$$

EXAMPLE 2.3
Determine the spectrum of the pulse shown in Figure 2.14.

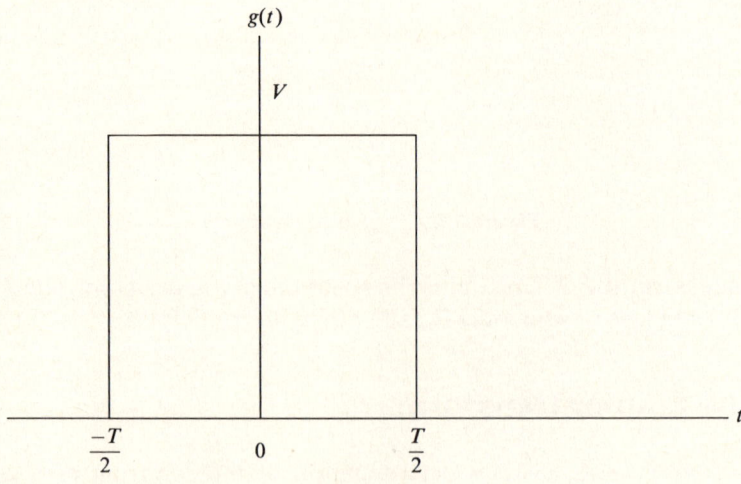

Figure 2.14

SOLUTION

First we shall define the pulse mathematically:

$$g(t) = 0, \qquad t < -T/2$$
$$= V, \qquad -T/2 \leqslant t \leqslant T/2$$
$$= 0, \qquad t > T/2$$

Substituting $g(t)$ into Eqn (2.25) defining a Fourier transform yields:

$$G(f) = V \int_{-T/2}^{T/2} e^{-j2\pi ft} \, dt \tag{2.26}$$

$$= \frac{-V}{j2\pi f} (e^{-j2\pi ft})_{-T/2}^{T/2} \tag{2.27}$$

$$= \frac{V}{j2\pi f} (e^{(j2\pi fT/2)} - e^{(-j2\pi fT/2)}) \tag{2.28}$$

$$= \frac{2V}{2\pi f} \left(\frac{e^{(j2\pi fT/2)} - e^{(-j2\pi fT/2)}}{j} 2 \right) \tag{2.29}$$

The contents of the bracket in Eqn (2.29) is of the general form shown in Eqn (2.30):

$$\sin \theta = \frac{e^{j\theta} - e^{-j\theta}}{j2} \tag{2.30}$$

Hence

$$\therefore \quad G(f) = \frac{2V}{2\pi f} \sin \left(\frac{2\pi fT}{2} \right) \tag{2.31}$$

$$= VT \left[\frac{\sin(2\pi fT/2)}{2\pi fT/2} \right] \tag{2.32}$$

$$= VT \left(\sin \frac{\pi fT}{\pi fT} \right) \tag{2.33}$$

The above equation, Eqn (2.33), for the spectrum of a pulse is a well-known expression which frequently occurs when using Fourier transforms and is called a **sinc** function. It is of the general form $\sin(x)/x$ but note when calculating that x is expressed in radian and not degree. We may therefore rewrite Eqn (2.33):

$$G(f) = VT \operatorname{sinc} fT \tag{2.34}$$

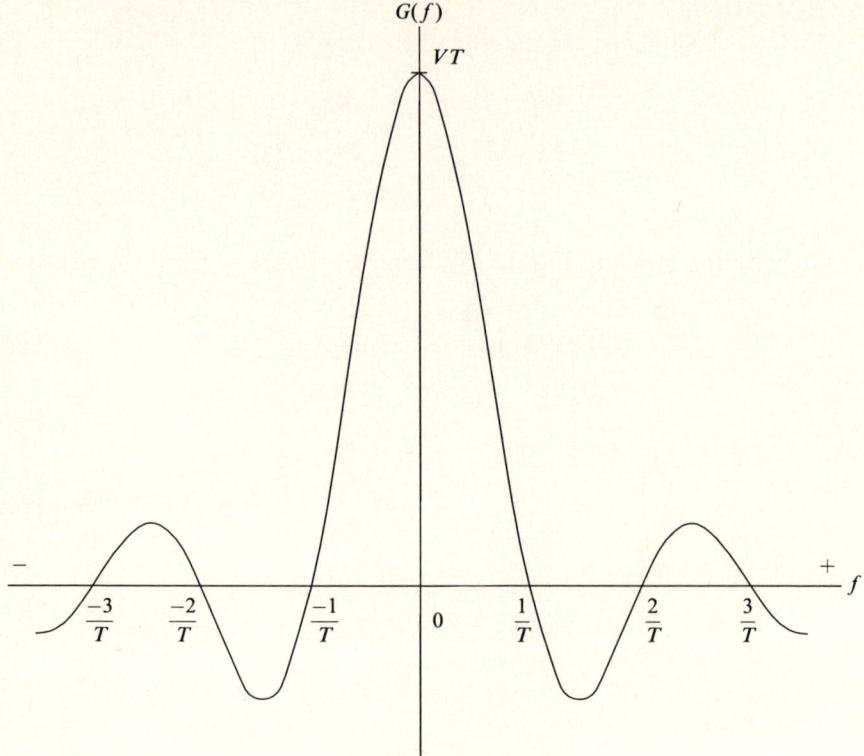

Figure 2.15 Spectrum of a single pulse

The spectrum is as shown in Figure 2.15 and reveals that a waveform that is not continuous over all time produces a spectrum that is continuous over all frequencies.

EXAMPLE 2.4
Determine the spectrum of the delta function shown in Figure 2.16.

SOLUTION
Here $g(t)$ only exists at $t = 0$. It is therefore only necessary to evaluate the Fourier transform at $t = 0$:

$$G(f) = \int_{-\infty}^{\infty} g(t)\,e^{-j2\pi ft}\,dt \tag{2.35}$$

$$= \int_{t=0} V\delta(t)\,e^{-j2\pi ft}\,dt \tag{2.36}$$

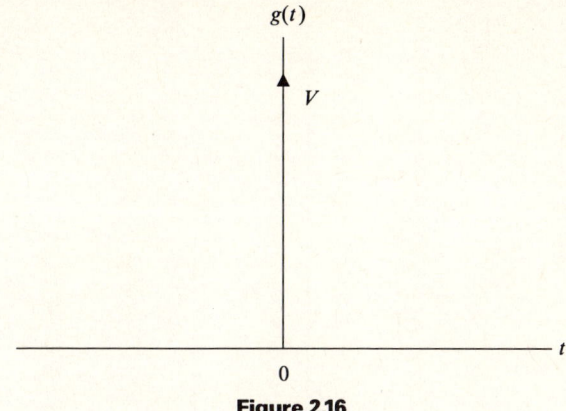

Figure 2.16

At $t = 0$, $\delta(t)$ is simply equal to 1. Hence at $t = 0$ Eqn (2.36) becomes:

$$G(f) = V \qquad (2.37)$$

This means that the spectrum of $V\delta(t)$ equals V for all frequency and is drawn in Figure 2.17.

This example is also useful to illustrate **duality**, or **reciprocity**. Suppose instead that $g(t)$ was a constant amplitude V for all time and that we are required to find the Fourier transform. In Example 2.4 we know that a constant amplitude V for all frequency inverse Fourier transforms to a delta function of amplitude V at $t = 0$. We may apply the concept of duality here: a constant for all time transforms to a delta function in the frequency domain.

More formally this process of reciprocity is set out in C.1.3 of Appendix C. We take the known spectrum $G(f)$ in Example 2.4, the constant for all frequency, and represent it as $G(t)$ with the frequency axis relabelled as time. We know that the original $G(f)$ inverse transforms to $g(t)$ and is a delta

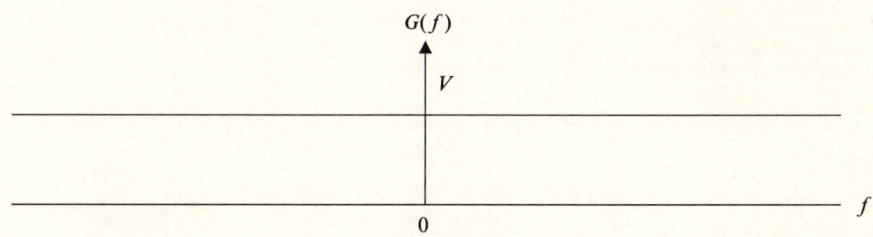

Figure 2.17

(a)

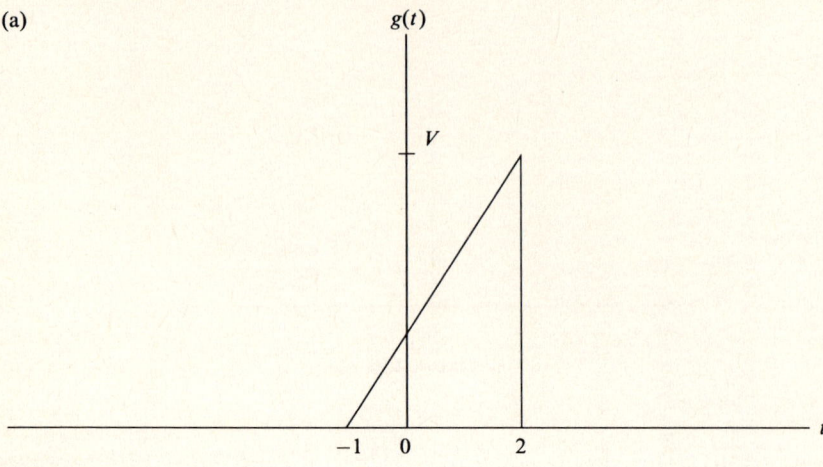

(b)

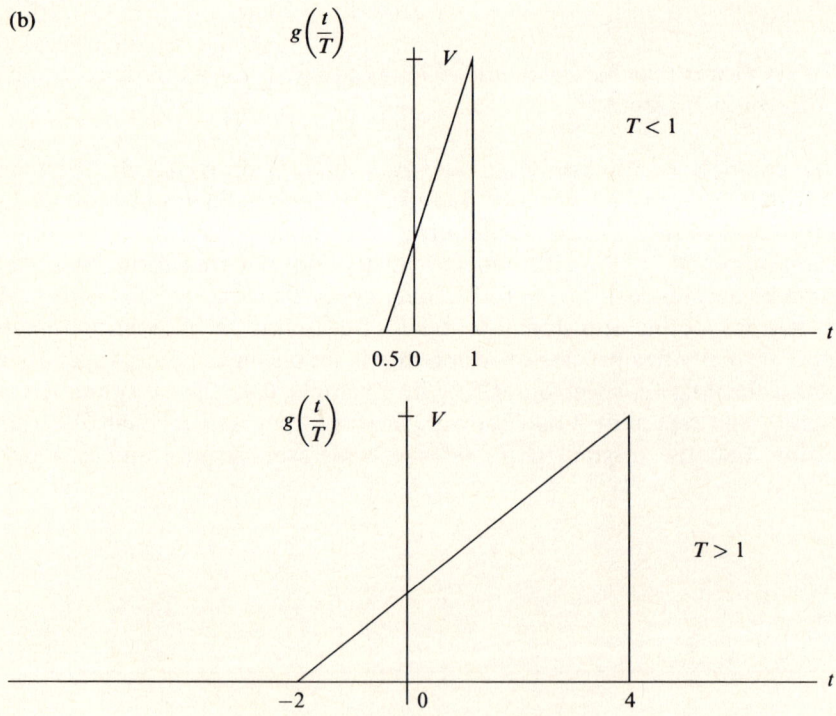

Figure 2.18   Time scaling: (a) $g(t)$; (b) $g(t/T)$

function. What Appendix C indicates is that our new expression $G(t)$ transforms to $g(-f)$ which is equivalent to $g(t)$ in Example 2.4, but with any event at time t now appearing upon a frequency axis at frequency $-f$. Simply, the function $g(t)$ in Example 2.4 now appears in the frequency domain with a reversal about the vertical axis. Hence, by means of reciprocity, we determine that a dc voltage V in the time domain transforms to $V\delta(f)$.

Self-assessment 2.3 A rectangular pulse has an amplitude of 1 and width of 1 second. If the pulse is centred upon time t_d, determine its Fourier transform and hence spectrum.

Suppose we have a signal $g(t)$ as shown in Figure 2.18(a). We may **time scale** this by the constant T such that $g'(t)$ becomes $g(t/T)$. Any event that occurred at $t = t_1$ in time now occurs at $T \times t_1$. Intuitively this must be so because, for a given event, the amplitude or value of $g(t/T)$ must equal $g(t)$. The effect of scaling by T is shown in Figure 2.18(b). If T is less than 1, $g(t)$ becomes compressed in time or, if the converse is the case, expanded. Where compression occurs in the time domain then, since events are occurring more rapidly, we may expect a corresponding expansion in the frequency domain, and vice versa.

EXAMPLE 2.5
The function $g(t)$ shown in Figure 2.19 is to be scaled by value 3. Sketch $g(t/3)$.

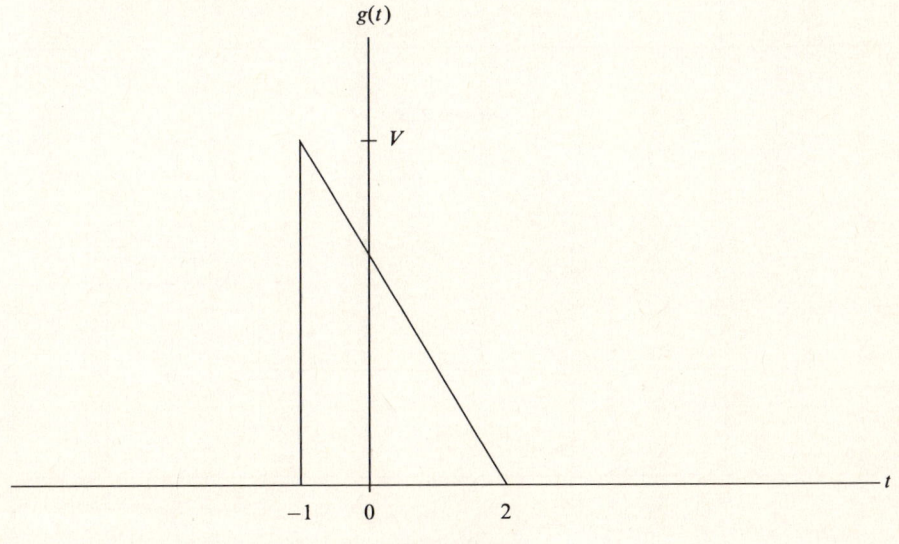

Figure 2.19

SOLUTION

As already indicated, dividing t by 3 has the effect of moving events at time t in $g(t)$ to time $3t$. Now, since T equals 3 this means that the waveform $g(t)$ is effectively 'stretched' by a factor of 3 as may be seen in Figure 2.20.

Self-assessment 2.4 The function $g(t)$ has the waveform shown in Figure 2.21. If $g(t)$ is time scaled to produce a new function $g(2t)$, state the value of scaling factor T and sketch the waveform of $g(2t)$.

As with Fourier series, we may save some effort in finding transforms because many well-known signals, and their spectra, have already been worked

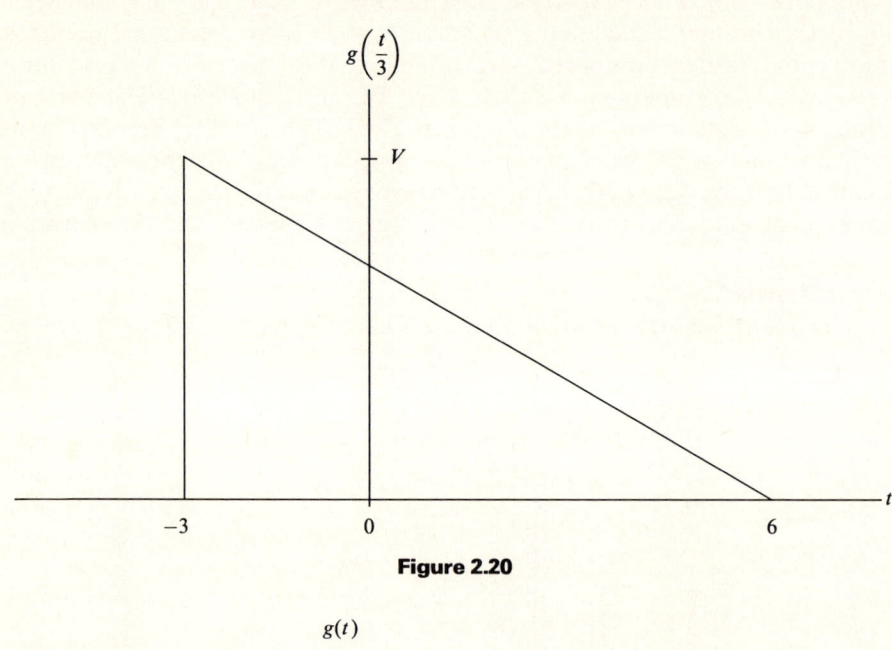

Figure 2.20

Figure 2.21

out for us and published in tables. Appendix C shows a number of Fourier transform pairs and their associated waveforms and spectra.

Now if we examine Appendix C we see that $g(t/T)$ transforms to $T \cdot G(fT)$. Note that in the frequency domain, in addition to frequency scaling, the amplitude is also scaled. As expected, if T is greater than 1, the effect of T in the frequency domain is to compress the spectrum. In order to retain the same area under the curve (which is a measure of energy contained in the signal), there is a corresponding expansion, by factor T, of the amplitude of $G(f)$.

Returning to Example 2.3, we may define the pulse shown in Figure 2.14 as a rectangular pulse with amplitude 1 and width 1 scaled by T. Hence $g(t)$ equals rect(t/T). From the tables we need to use the Fourier transforms of a rect(t) in conjunction with that of scaling by $g(t/T)$.

If $g(t)$ equals rect(t), then from tables $G(f)$ equals sinc(f).

Scaling by T: $g(t/T)$, from tables, becomes $T \cdot G(fT)$. That is $G(f)$ is amplitude scaled by factor T and wherever f appears in the expression for $G(f)$, it is replaced by fT.

Thus we may write:

$$g(t) = \text{rect}(t/T) \rightleftharpoons T \cdot G(fT) \tag{2.38}$$

$$= T\text{sinc}(fT) \tag{2.39}$$

The transform of rect(t/T) is found by tables to produce the same spectrum as that found in Example 2.3 by means of first principles. This example illustrates how tables may be used to find the spectrum of a signal which may consist of some basic wave shape, or function $g(t)$, and upon which one, or more, operations are performed. In this case we saw how the spectrum of $g(t)$, operated upon by time scaling, is obtained purely by systematically working through standard tables.

Another operation which may be applied to a signal $g(t)$ is that of **time shifting**. Suppose the rectangular pulse of Figure 2.14 is moved positively in time by T as shown in Figure 2.22. Then, as in scaling, $g(t)$ becomes $g'(t)$ where:

$$g'(t) = g(t - T)$$

That is any event that occurred at t equal to t_1 in time now occurs at interval $t_1 + T$. This must be so because, for a given event, the amplitude of $g(t - T)$ at time t must now equal $g(t)$.

Self-assessment 2.5 A waveform $g(t)$ is as shown in Figure 2.23. It is time shifted by -5 such that $g'(t) = g(t + 5)$.

(a)   Sketch $g'(t)$.
(b)   Use Fourier transform tables to find $G'(f)$.

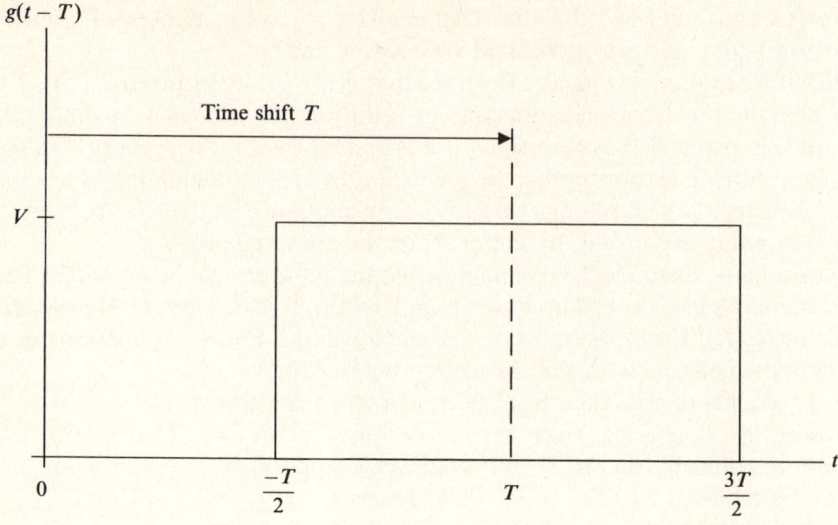

Figure 2.22   Time shift

(c)   Compare your answer in (b) above using tables with that obtained by first principles using the Fourier transform.

As already indicated, multiple operations may be performed upon a signal. Consider Example 2.6.

EXAMPLE 2.6

Given that $g(t)$ is as shown in Figure 2.24(a) determine the waveform of the function $g(t/3 + 2)$.

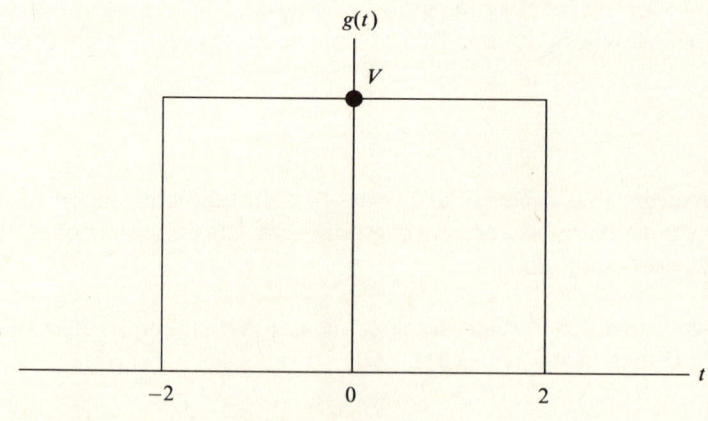

Figure 2.23

(a)

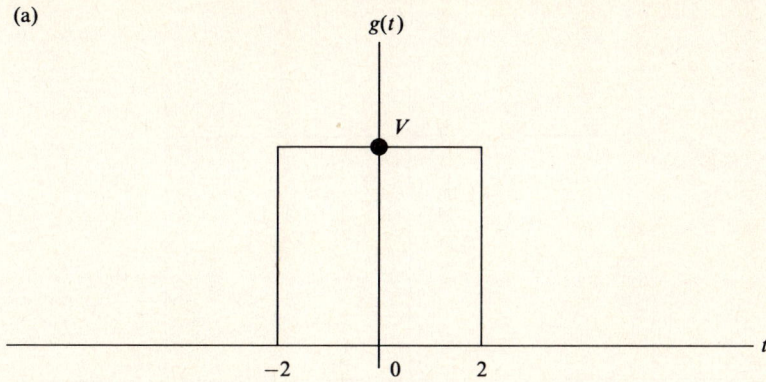

(b)

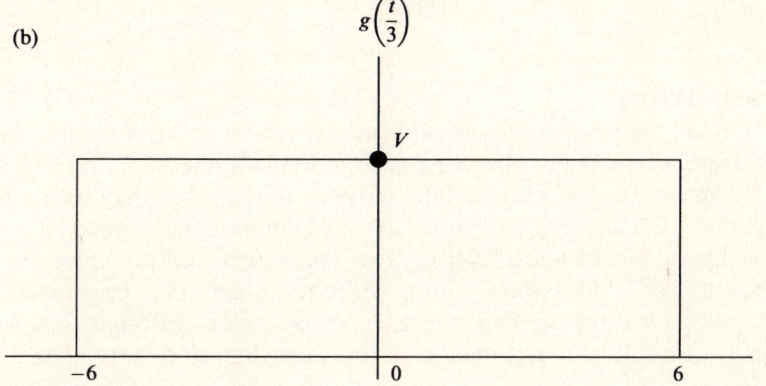

(c)

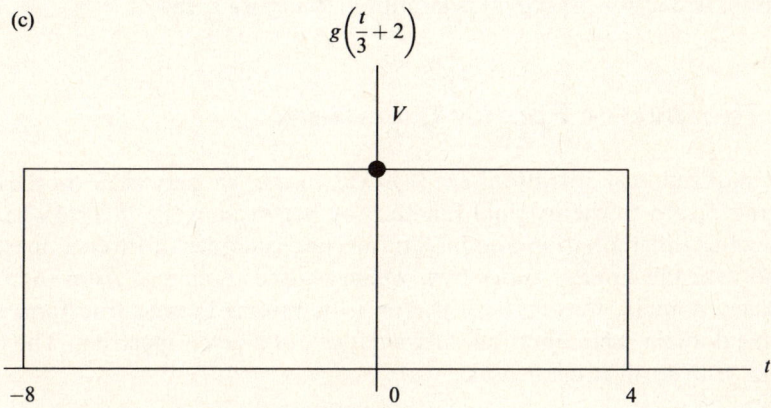

Figure 2.24

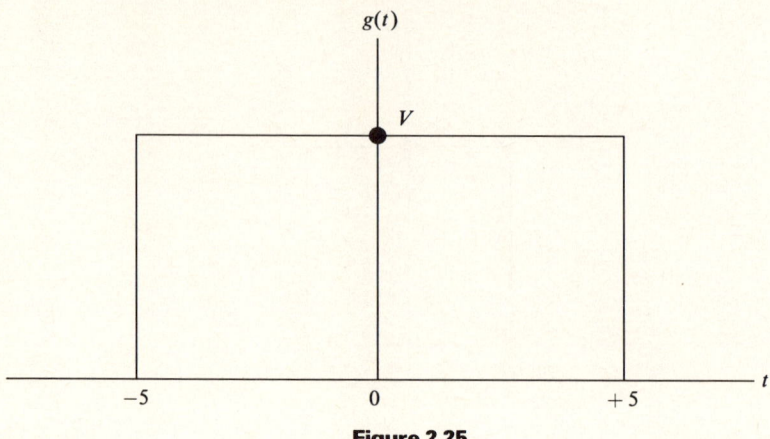

Figure 2.25

SOLUTION

With a little practice it is possible in many cases to write the new function directly by carefully observing each individual operation applied to $g(t)$. However, we shall move a little more cautiously. We may see that $g(t)$ is scaled by 3 and shifted (to the left) by 2 units. Firstly we shall apply scaling to yield Figure 2.24(b). Now applying a shift of $+2$ we obtain the result $g(t/3 + 2)$, Figure 2.24(c). Since these are linear operations we could have done the two stages in reverse order, shift and then scaling, but the result would be identical. Try it for yourself by way of an exercise.

Self-assessment 2.6 Determine, by means of Fourier transform tables, the spectrum of $g(t/4 - 3)$ if $g(t)$ is as shown in Figure 2.25.

2.4 The inverse Fourier transform

Most mathematical operations are two-way in that we may perform the operation and return to the original function by performing the inverse operation. In calculus, differentiation and integration are examples of inverse operations. So too with the Fourier transform, which is used to change from the time to frequency domain, we may also perform the **inverse Fourier transform** to find the time domain representation, or waveform, of a given spectrum. The inverse Fourier transform is defined as:

$$g(t) = \int_{-\infty}^{\infty} G(f)\,e^{j2\pi ft}\,df \tag{2.40}$$

EXAMPLE 2.7

Use the inverse Fourier transform to determine the time domain signal representation of $VT\,\text{sinc}(fT)$.

$$g(t) = VT \int_{-\infty}^{\infty} \frac{\sin(\pi fT)}{\pi fT}\, e^{j2\pi ft}\, df \tag{2.41}$$

SOLUTION

Substituting the following into Eqn (2.41):

$$e^{j2\pi ft} = \cos 2\pi ft + j \sin 2\pi ft \tag{2.42}$$

yields:

$$g(t) = \frac{V}{\pi} \left(\int_{-\infty}^{\infty} \frac{\sin \pi ft \cos 2\pi ft}{f}\, df + j \int_{-\infty}^{\infty} \frac{\sin \pi ft \sin 2\pi ft}{f}\, df \right) \tag{2.43}$$

The right hand integral of Eqn (2.43) is the product of two odd functions which is therefore also odd. The integral of an odd function over all time is zero. Hence the right hand integral may be ignored. Rewriting Eqn (2.43) and noting that the left hand integral is even which may be rewritten as twice the integral with limits 0 to $+\infty$:

$$\therefore \quad g(t) = \frac{2V}{\pi} \int_{0}^{\infty} \frac{\sin \pi ft \cos 2\pi ft\, df}{f} \tag{2.44}$$

Now:

$$\sin \pi ft \cos 2\pi ft = \tfrac{1}{2}[\sin(2\pi ft + \pi ft) - \sin(2\pi ft - \pi ft)] \tag{2.45}$$

$$\therefore \quad g(t) = \frac{V}{\pi} \int_{0}^{\infty} \frac{\sin[\pi(2t + T)f]}{f}\, df - \int_{0}^{\infty} \frac{\sin[\pi(2t - T)f]}{f}\, df \tag{2.46}$$

Using the standard integral:

$$\int_{0}^{\infty} \frac{\sin ax}{x}\, dx = \frac{\pi}{2}, \qquad a > 0$$

$$0, \qquad a = 0$$

$$-\frac{\pi}{2}, \qquad a < 0$$

Let the left hand integral of Eqn (2.46) be represented by I_1 and that of the right hand by I_2, then:

$$g(t) = \frac{V}{\pi}(I_1 - I_2) \tag{2.47}$$

where:

$$-\infty < t < T/2 \quad \text{gives} \quad I_1 = -\pi/2,\; I_2 = -\pi/2, \qquad \therefore \quad I_1 - I_2 = 0$$

$$-T/2 < t < T/2 \quad \text{gives} \quad I_1 = \pi/2, \quad I_2 = -\pi/2, \qquad \therefore \quad I_1 - I_2 = \pi$$

$$T/2 < t < \infty \quad \text{gives} \quad I_1 = \pi/2, \quad I_2 = \pi/2, \qquad \therefore \quad I_1 - I_2 = 0$$

Hence $g(t)$ in Eqn (2.47) becomes:

$$g(t) = \frac{V}{\pi} \pi = V, \quad \text{for} \quad -\frac{T}{2} < t < \frac{T}{2} \tag{2.48}$$

Hence we see that the spectrum of $VT\,\mathrm{sinc}(fT)$ inverse transforms to $V\,\mathrm{rect}(t/T)$. In Example 2.3 we found the Fourier transform of $V\,\mathrm{rect}(t/T)$ and here we now see that, as expected, the inverse Fourier transform $VT\,\mathrm{sinc}(fT)$ is the same.

2.5 Convolution

We can, in principle, apply any mathematical operation to a signal. We have already seen how, by means of integration, it is possible to move between frequency and time domains. Common operations performed upon signals are addition, subtraction, multiplication and division (although the latter may alternatively be regarded as multiplying by a number less than 1).

With reference to the communications model shown in Chapter 1 we see that the channel has associated with it a transfer function which simply represents the mathematical expression relating output $K(f)$ to the input $G(f)$. For example, if a signal represented by $G(f)$ is applied to the channel, then the output $K(f)$ is expressed:

$$K(f) = G(f) \cdot H(f) \tag{2.49}$$

Now $k(t)$ may then be found by taking the inverse Fourier transform of $K(f)$. However, in some cases either $G(f)$ or $H(f)$ may not be readily expressed, or their product may prove to be difficult. Then, as an alternative, $k(t)$ may often more easily be found in a different manner shown:

$$k(t) = \mathcal{F}^{-1}[G(f) \cdot H(f)] \tag{2.50}$$

$$= \int_{-\infty}^{\infty} G(f) \cdot H(f)\, e^{j2\pi ft}\, df \tag{2.51}$$

where $\mathcal{F}^{-1}$ is mathematical shorthand for 'inverse Fourier transform'.

Integration of the product in Eqn (2.51) above, where there are two functions of f, may be achieved by the introduction of a dummy variable, say u. Then, by expressing $G(f)$ in the form of a Fourier transform, we may write:

$$k(t) = \int_{-\infty}^{\infty} [g(u)\,e^{-j2\pi fu}\,du] H(f)\,e^{j2\pi ft}\,df \qquad (2.52)$$

Interchanging the order in which we integrate:

$$k(t) = \int_{-\infty}^{\infty} g(u)\left[\int_{-\infty}^{\infty} H(f)\,e^{j2\pi(t-u)f}\,df\right]du \qquad (2.53)$$

Now if we compare Eqn (2.53) with that of (2.40) we see that the content of the square bracket is simply the inverse Fourier transform of $H(f)$ time shifted by u. Hence we may write:

$$H(f)\,e^{-j2\pi uf} \rightleftharpoons h(t - u) \qquad (2.54)$$

$$\therefore \quad k(t) = \int_{-\infty}^{\infty} g(u) \cdot h(t - u)\,du \qquad (2.55)$$

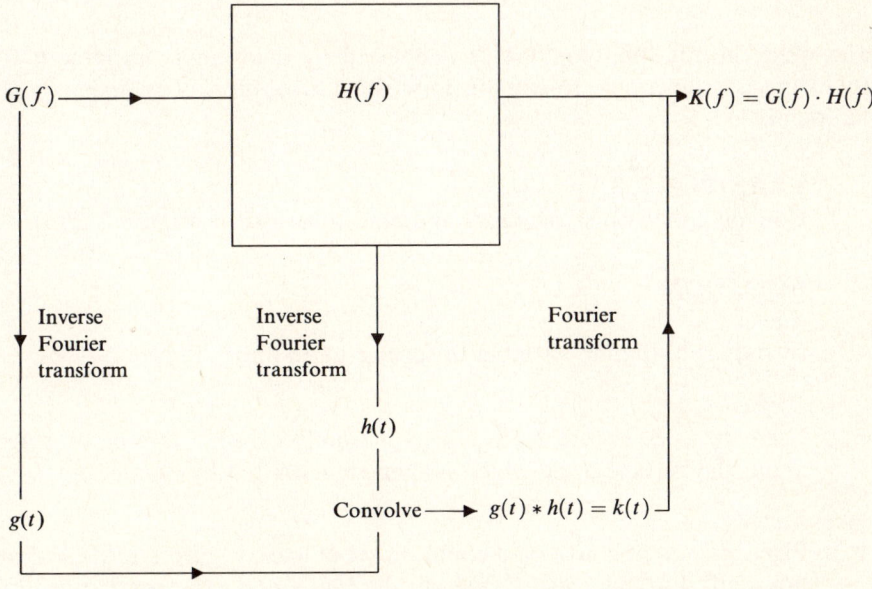

Figure 2.26 Convolution

Equation (2.55) is the mathematical operation known as **convolution**. Thus we see that multiplication in the frequency domain is convolution in the time domain. (This is commutative where the converse is true – convolution in the frequency domain is equivalent to multiplication in the time domain.) Mathematically, convolution is indicated by an asterisk rather than the × of multiplication. Figure 2.26 illustrates the use of convolution, as an alternative operation to multiplication.

It can be seen that instead of multiplying frequency responses to obtain $K(f)$ we may take the inverse Fourier transform of both $G(f)$ and $H(f)$. They may then be convolved to form $k(t)$. Finally to obtain the spectrum $K(f)$ we take the Fourier transform of $k(t)$.

Convolution and multiplication are complementary, but not identical, mathematical operations and their interdependence is shown:

$$G(f) \cdot H(f) = g(t) * h(t) \tag{2.56}$$

$$g(t) \cdot h(t) = G(f) * H(f) \tag{2.57}$$

Convolution may be defined:

$$g(t) * h(t) = \int_{-\infty}^{\infty} g(u)h(t-u)\,\mathrm{d}u \tag{2.58}$$

$$= \int_{-\infty}^{\infty} g(t-u)h(t)\,\mathrm{d}u \tag{2.59}$$

and where $g(u)$ is simply $g(t)$ with t replaced by the dummy variable u. The following is an example to explain how the convolution of two signals may be found.

EXAMPLE 2.8
Convolve the two signals $g(t)$ and $h(t)$ illustrated in Figure 2.27(a).

SOLUTION
Step 1
Introduce a dummy variable to form $g(u)$ and $h(u)$, Figure 2.27(b).

Step 2
Form $g(t-u)$, Figure 2.27(c). This is simply $g(-u)$, a folding of $g(u)$ about the vertical axes, which is then right-shifted by an amount t.

Step 3
Place $g(t-u)$ and $h(u)$ on a common set of axes, Figure 2.27(d). We may now shift, or 'slide', $g(t-u)$ along the horizontal axis from a position $u = -\infty$ to $u = +\infty$ by suitable variation in t. Careful inspection of

(a)

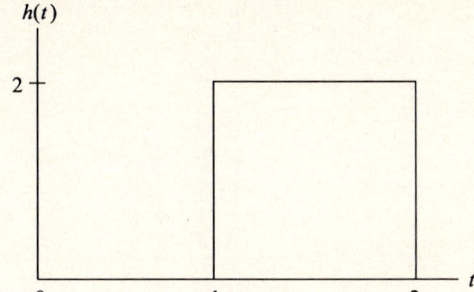

(b)

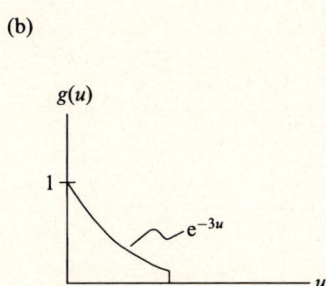
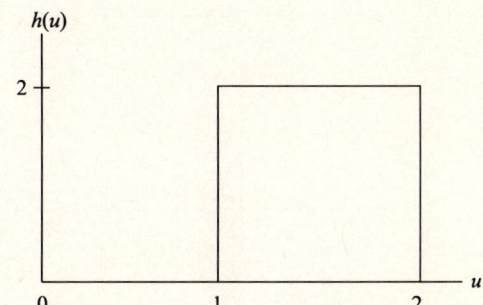

(c)

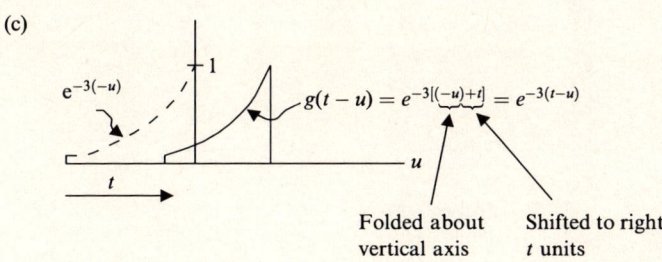

$$g(t - u) = e^{-3[(-u)+t]} = e^{-3(t-u)}$$

Folded about vertical axis   Shifted to right t units

(d)

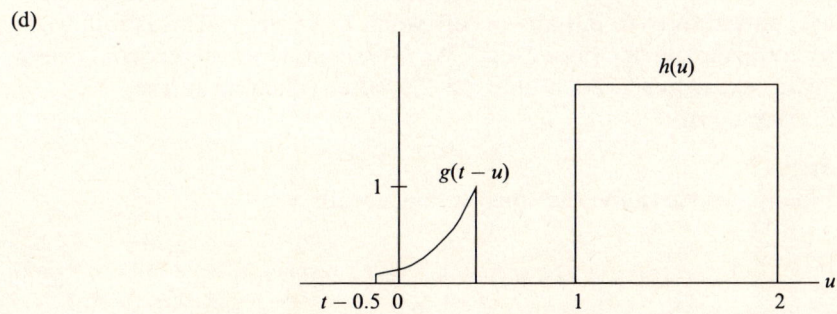

(e)

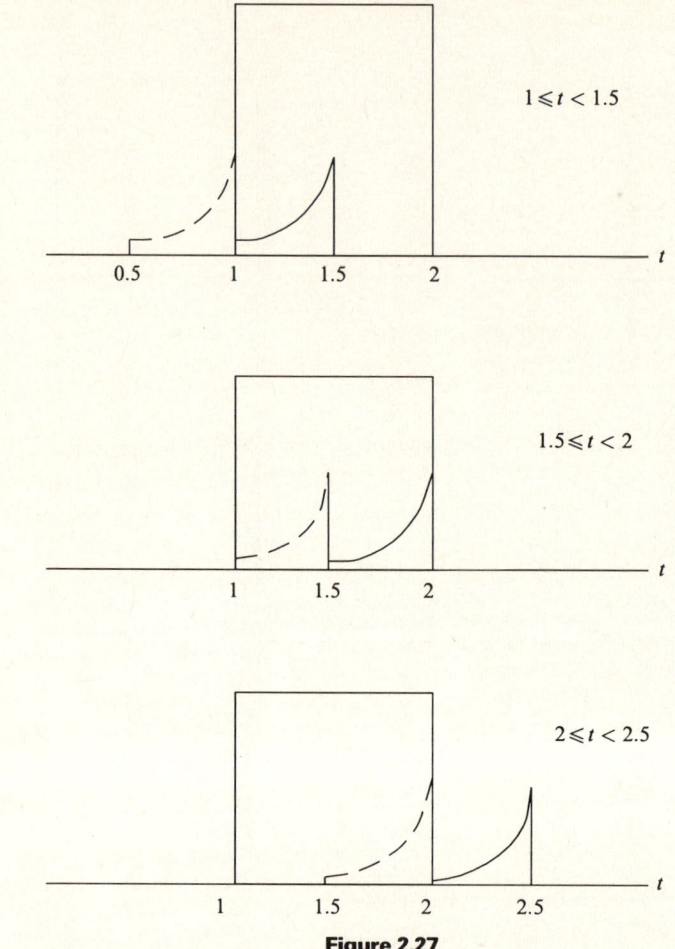

$1 \leqslant t < 1.5$

$1.5 \leqslant t < 2$

$2 \leqslant t < 2.5$

Figure 2.27

Figure 2.27(d) reveals that there are three distinct regimes which are indicated in Figure 2.27(e). As $g(t - u)$ is gradually slid from the left, partial overlap initially occurs when t exceeds 1. This is followed by complete overlap once $t \geqslant 1.5$. As t is increased further, partial overlap again occurs, when $2 \leqslant t$, until $t > 2.5$, after which no further overlap occurs.

Step 4
Finally, to determine the convolution, recall:

$$g(t) * h(t) = \int_{-\infty}^{\infty} g(t - u)h(t)\,du \qquad (2.60)$$

The convolution integral is the area under the curve of the product $g(t - u)$ and $h(u)$. Clearly this is zero for $t < 1$ or $t > 2.5$ because outside the range $1 \leqslant t < 2.5$ there is no overlap and hence the product is zero. Thus, by examining Figure 2.27(e) carefully, we may write:

$$g(t) * h(t) = \int_1^t e^{-3(t-u)} 2du, \qquad 1 \leqslant t < 1.5 \qquad (2.61)$$

$$= \int_{t-0.5}^t e^{-3(t-u)} 2du, \qquad 1.5 \leqslant t < 2 \qquad (2.62)$$

$$= \int_{t-0.5}^2 e^{-3(t-u)} 2du, \qquad 2 \leqslant t < 2.5 \qquad (2.63)$$

Note how in Eqn (2.61) integration commences from $t = 1$, the integral equalling zero for $t < 1$. In Eqn (2.62), the lower limit of the integral begins at the position $t - 0.5$, that is the left hand edge of $g(t - u)$, and must equal 1 when $t = 1.5$. This limit corresponds with the left hand edge of $g(t - u)$ progressively moving from $u = 1$ to $u = 1.5$ as $g(t - u)$ slides to the right. The limit progressively cuts off the lower limit of integration as $g(t - u)$ slides to the right. The upper limit of integration is set by the range limit of the integration $t = 2$. Finally, Eqn (2.63), the left hand edge of $g(t - u)$ cuts off the integration from $u = 1.5$ through 2, the latter terminating the upper limit of integration for all u. Hence, as before, the lower limit of integration is $t - 0.5$ and the upper limit is bounded by $t = 2$.

We may now complete integration in Eqns (2.61), (2.62) and (2.63) to yield the result of convolving $g(t)$ and $h(t)$:

$$g(t) * h(t) = \tfrac{2}{3} e^{-3t}(e^{3u})_1^t \qquad 1 \leqslant t < 1.5 \qquad (2.64)$$

$$= \tfrac{2}{3} e^{-3t}(e^{3u})_{t-0.5}^t \qquad 1.5 \leqslant t < 2 \qquad (2.65)$$

$$= \tfrac{2}{3} e^{-3t}(e^{3u})_{t-0.5}^2 \qquad 2 \leqslant t < 2.5 \qquad (2.66)$$

Hence:

$$g(t) * h(t) = \tfrac{2}{3}(1 - e^{-3(t-1)}) \qquad 1 \leqslant t < 1.5 \qquad (2.67)$$

$$= \tfrac{2}{3} e^{-3t}(e^{3t} - e^{3t} e^{1.5}) \qquad (2.68)$$

$$= 0.5179 \qquad 1.5 \leqslant t < 2 \qquad (2.69)$$

$$= \tfrac{2}{3} e^{-1.5}(e^{-(3t-7.5)} - 1) \qquad 2 \leqslant t < 2.5 \qquad (2.70)$$

The results shown in Eqns (2.66) and (2.67) may be sketched to show pictorially the effect, in Figure 2.28, of convolving $g(t)$ and $h(t)$.

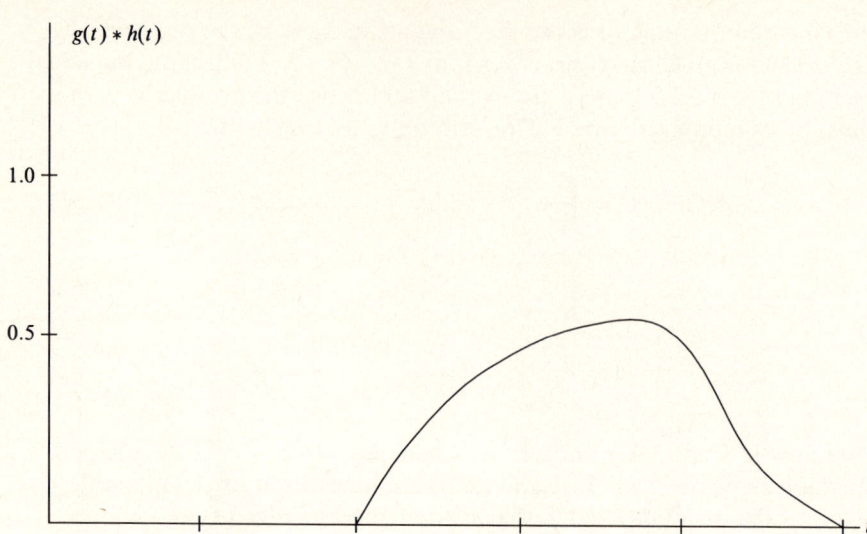

Figure 2.28 Sketch of $g(t)$ and $h(t)$ convolved

In the above example $g(t)$ was deliberately chosen to have enough variation to highlight the essential points in general to master when convolving. Many simpler problems can, after some practice of other convolution problems, be done purely by means of sketches with no recourse to piece-wise integration as in the example, and which can be rather laborious. It is apparent from Figure 2.28 that the convolved signal ranges from $t = 1$ to $t = 2.5$. That is, the range now equals the sum of the ranges of $g(t)$ and $h(t)$. This effect always occurs and is often termed 'spreading' or 'smearing'. Note also that, although the calculations differ, exactly the same result would have occurred had we decided to fold and slide $h(t)$ rather than using $g(t - u)$. It is suggested as an exercise that Example 2.8 be repeated in the above manner. Convolution may be viewed pictorially as a summing of the overlap area as one function slides across the other in the convolution operation.

Self-assessment 2.7 Convolve the two signals shown in Figure 2.29 both by integration and directly using graphs.

2.6 Analogue filters

A filter is merely a circuit that places a limit upon the range of frequencies it will pass, known as **passband**, and rejects any frequencies that fall outside this range. There are four basic filter responses, namely low pass, high pass, band pass and band stop (or notch). Their idealised and more typical practical

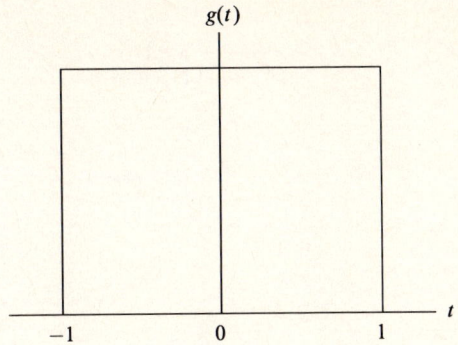

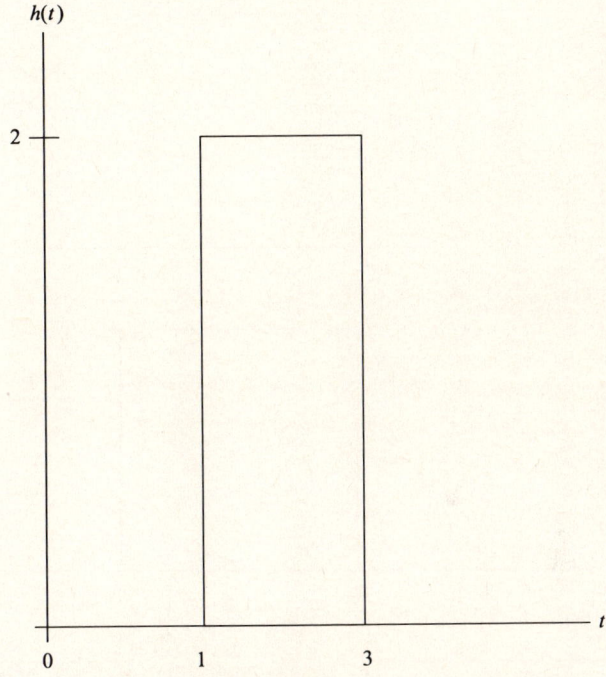

Figure 2.29

responses are shown in Figure 2.30, where lower and upper cut-off frequencies are represented by f_l and f_u, respectively. In practice the sharpness of cut-off curves and also the degree to which attenuation remains constant within the pass band is at variance to that of the ideal.

Filters are commonly found in communication systems. Their purposes are varied but include selection of a single channel within a multi-channel system

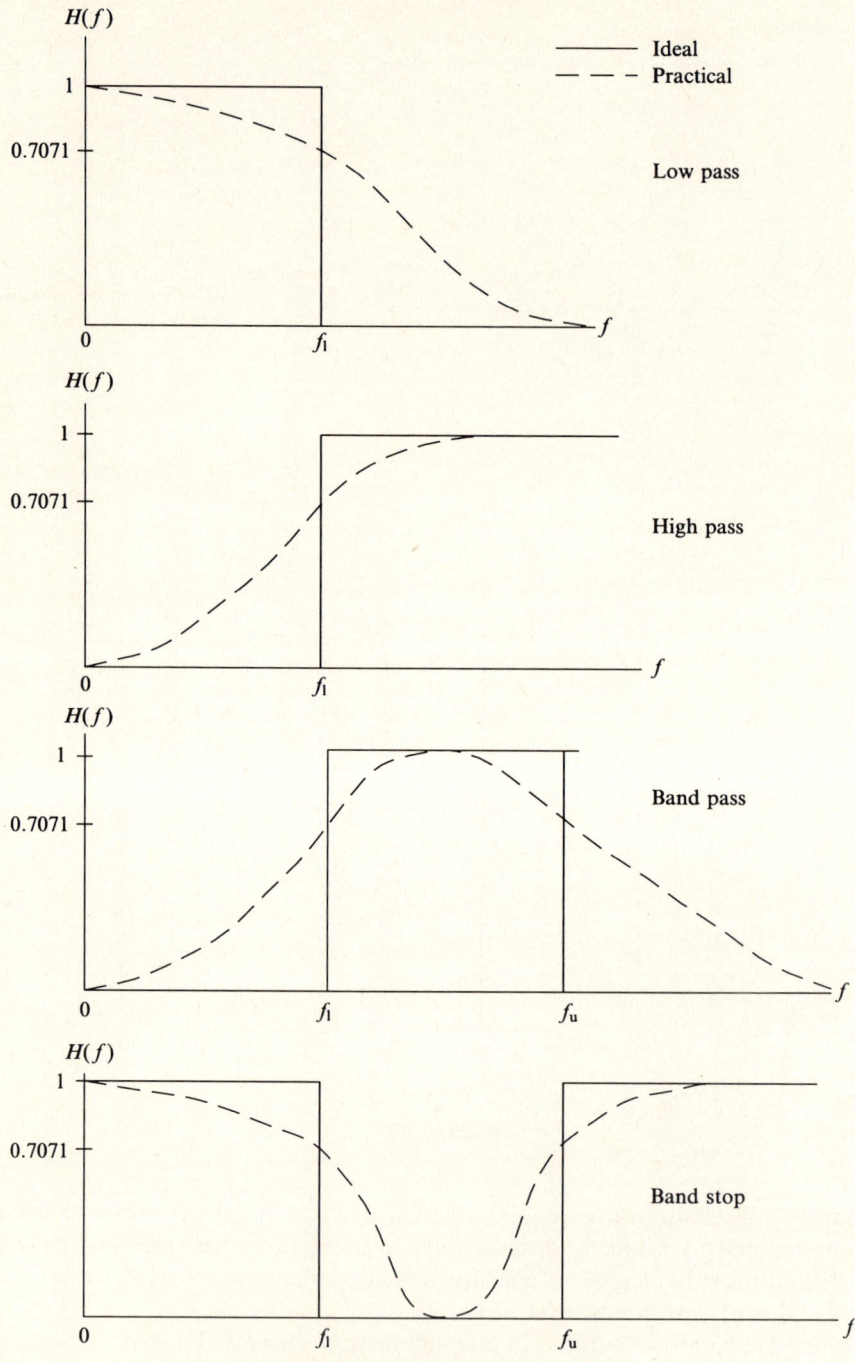

Figure 2.30 Idealised filter responses

where channels occupy contiguous bands of frequency. Another purpose is to limit the input to a receiver to the frequency range of the expected signal. In this way any noise and/or interference at nearby frequencies are rejected to improve the overall quality of the system.

Filters may be **passive** in that they do not contain any active elements, e.g. transistors or integrated circuits. Two simple examples are shown in Figure 2.31 with their respective transfer functions which illustrate how practical filter responses deviate from that of the ideal.

Filters that contain active elements, usually in the form of operational amplifiers, are called **active filters**. The main advantage of an active filter is that all four types of filter may be realised with the aid of resistive and capacitive passive elements, there being no necessity to use inductor coils. In consequence

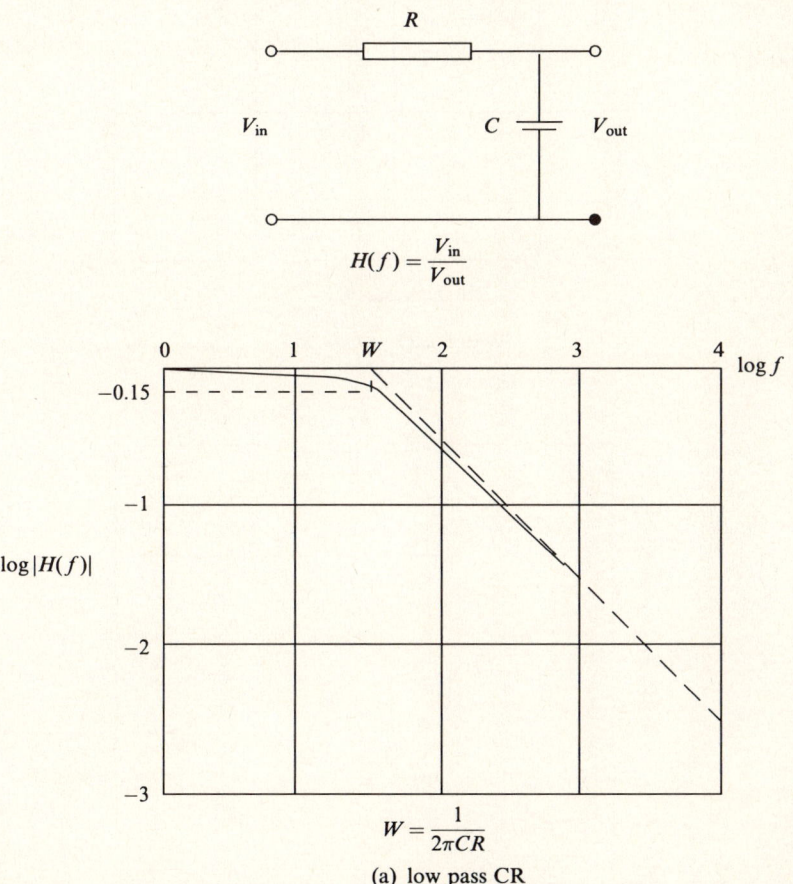

$$H(f) = \frac{V_{in}}{V_{out}}$$

$$W = \frac{1}{2\pi CR}$$

(a) low pass CR

Figure 2.31   Simple filters (passive)

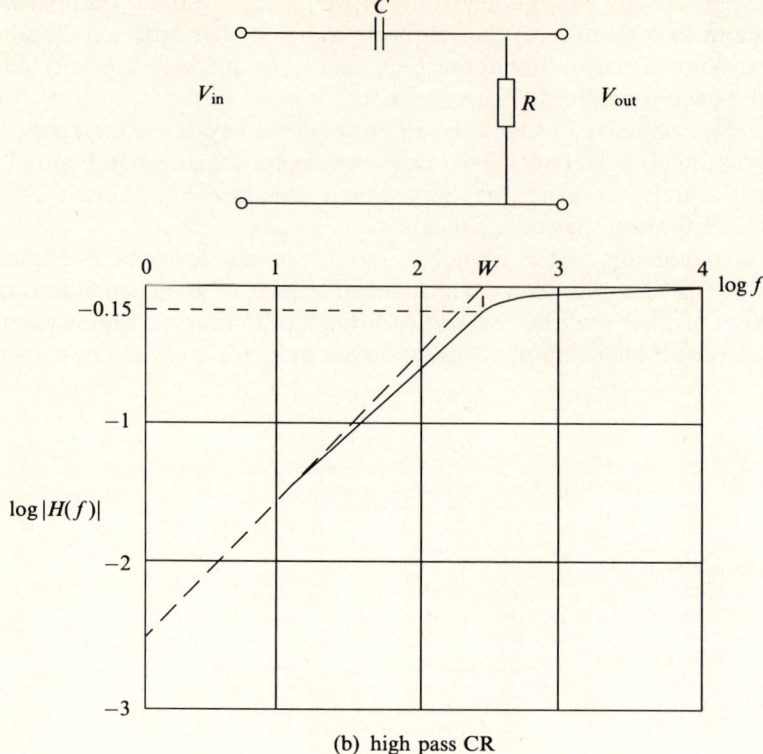

(b) high pass CR

Figure 2.31 *Continued*

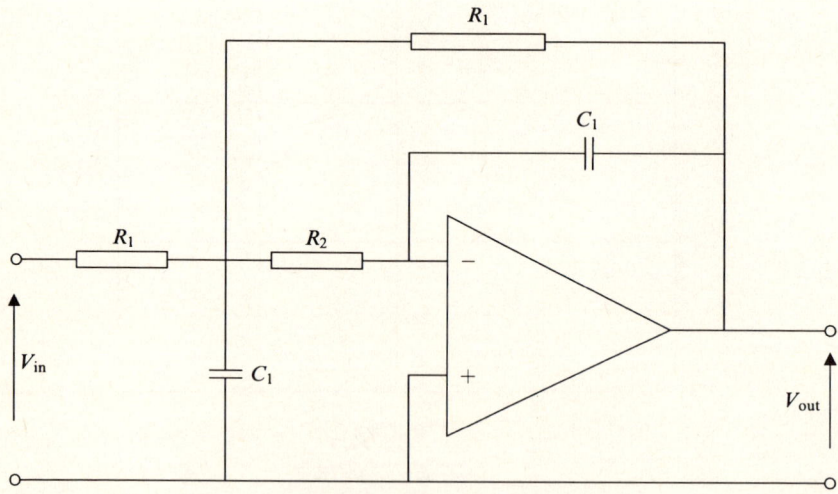

Figure 2.32 An active low-pass filter

active filters may be produced which are physically small and light; they are capable of being fabricated into integrated circuit packages. A typical example of an active filter is shown in Figure 2.32.

2.7 An introduction to digital signal processing

Digital signal processing (DSP) samples signals periodically, usually by means of an ADC. The samples therefore become a series of numbers which may then be processed using computer-based techniques. Signal processing performed using analogue techniques, by means of electronic circuits, may be performed equally well digitally using digital signals derived via an ADC, providing certain conditions are met which are discussed in the following section. DSP is very attractive since it is performed in software, making it convenient and simple to use. DSP is often cheaper, more reliable and precise than analogue alternatives.

2.7.1 Sampling

Suppose we sample an arbitrary waveform $g(t)$ periodically every T seconds. We shall assume a perfect sampling operation whereby $g(t)$ is multiplied by a sampling signal $s(t)$. In practice $G(f)$, the Fourier transform of $g(t)$, is band-limited to W hertz.

The sampling signal $s(t)$ may be represented by the following shorthand:

$$s(t) = \text{rep}_T[\delta(t)] \tag{2.71}$$

This means that $s(t)$ is a function which is repeated at interval T. In this case the function that is being repeated is a delta function. Sampling in the time domain produces $k(t)$, which is the product of $g(t)$ and $s(t)$:

$$k(t) = g(t) \cdot s(t) \tag{2.72}$$

$$= g(t) \cdot \text{rep}_T[\delta(t)] \tag{2.73}$$

To find the spectrum $K(f)$ we may convolve $G(f)$ and the Fourier transform of $s(t)$. From C.2.6 in Appendix C we may note the Fourier transform pair:

$$\text{rep}_T[g(t)] \leftrightarrow G(f) \cdot \frac{1}{T} \text{rep}_{1/T}[\delta(f)] \tag{2.74}$$

From Eqn (2.71), noting that the Fourier transform of the delta function is 1, the Fourier transform of $s(t)$, with reference to Eqn (2.74), may be written:

$$S(f) = 1 \cdot \frac{1}{T} \text{rep}_{1/T}[\delta(f)] \tag{2.75}$$

Hence $K(f)$ is:

$$K(f) = G(f) * \frac{1}{T} \text{rep}_{1/T}[\delta(f)] \tag{2.76}$$

The spectrum of $K(f)$ is as shown in Figure 2.33 where we may see that the original spectrum $G(f)$ is repeated in the frequency domain at intervals of $1/T$ to $+\infty$ Hz. In order to transmit such a signal only one occurrence of $G(f)$ is necessary. This may be achieved by passing $K(f)$ through a low-pass filter with a cut-off frequency of W Hz. This will pass one complete occurrence of $G(f)$ centred upon 0 Hz and reject all other repetitions of $G(f)$ in the spectrum of $K(f)$.

There is a limit upon the sampling rate f_s of $s(t)$. If the rate is so low that $1/T$ is reduced to a point where $1/T \leqslant 2W$, the repetitions of $G(f)$ overlap as shown in Figure 2.34.

Overlap of adjacent spectral repetitions cause the higher frequency elements of one spectral repetition to combine with lower frequency elements of an adjacent repetition. That is any one spectral occurrence no longer merely contains the original baseband signal, but also interfering frequency elements from an adjacent spectral repetition. This phenomenon is known as **aliasing**. Once aliasing occurs the original signal is corrupted and is therefore no longer recoverable in its original form. In order to prevent aliasing the **Nyquist criterion** must be satisfied. That is the sampling frequency must be at least twice the highest frequency component of the signal being sampled, or:

$$f_s \geqslant 2W \tag{2.77}$$

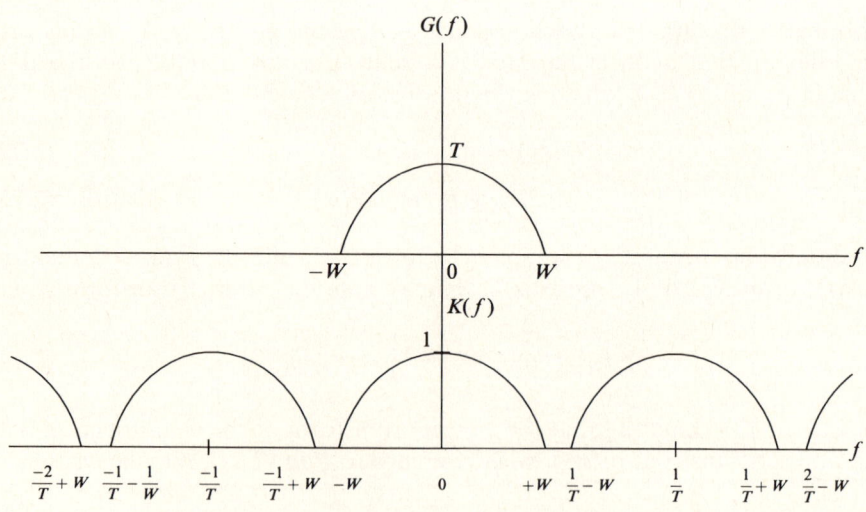

Figure 2.33   Spectrum of a sampled signal

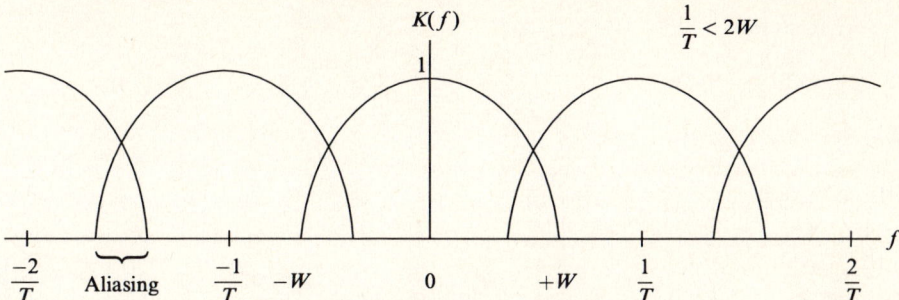

Figure 2.34 Effect of sampling rate too low

2.7.2 A number sequence signal representation

We saw in the last section how a signal $g(t)$ may be periodically sampled at interval T. We may represent a sampled signal by $g(n)$ as shown in Figure 2.35. The sampled signal $g(n)$ is merely a succession of $g(t)$ amplitudes, each at successive intervals of n and, since the intervals are spaced T seconds apart in time, may be thought of as a time signal. For DSP purposes we may simply regard $g(n)$ as a series of values, or numbers, which are known as a **number sequence**. Providing that the sampling interval T is sufficiently short to satisfy the Nyquist criterion then, as we saw in Section 2.7.1, the number sequence contains all of the original information contained in the signal $g(t)$. Some standard digital signals represented as number sequences are shown in Figure 2.36.

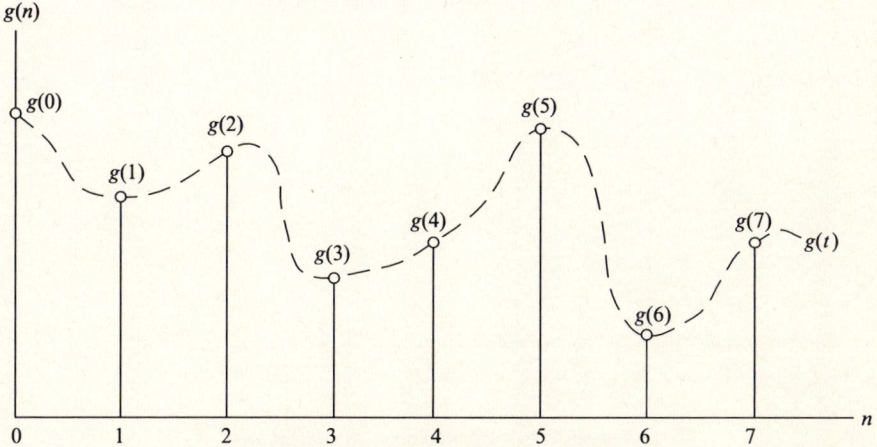

Figure 2.35 Signal as a number sequence

(a) Unit step

$$u(n) = 0, \quad n < 0$$
$$= 1, \quad \geqslant 1$$

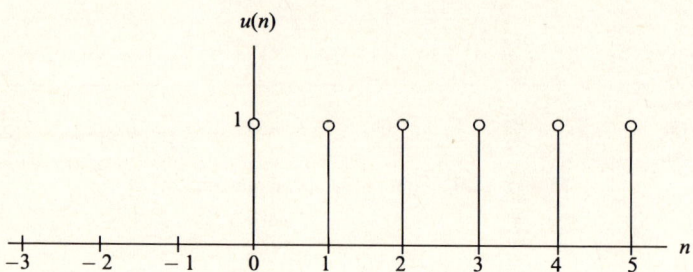

(b) Ramp

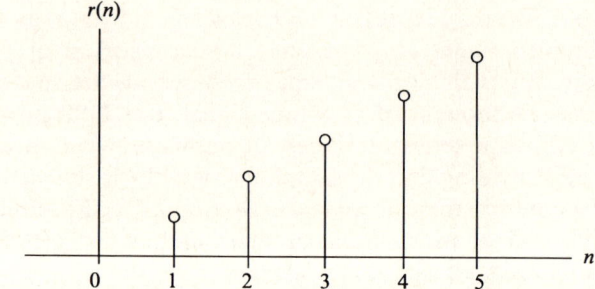

$r(n) = kv(n)$ where k is a
constant and determines gradient

(c) Delta

$$f(n) = \delta(n + 2)$$

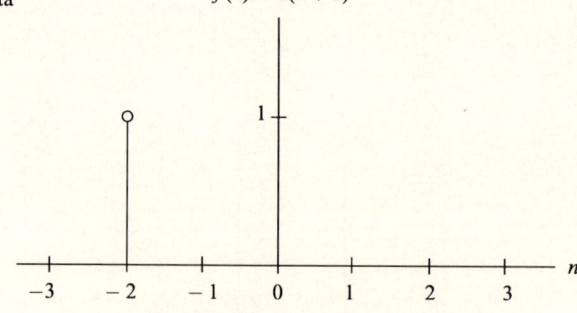

Figure 2.36 Digital signal representation

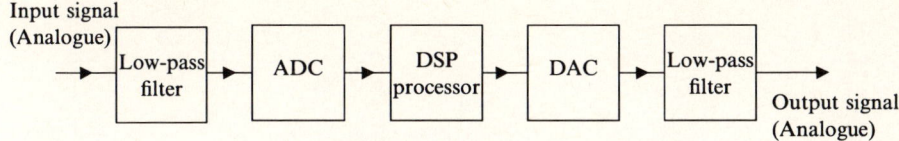

Figure 2.37   A DSP system

Having established an alternative signal representation in the form of a number sequence it is possible to perform digital signal processing with a computer, or by means of a dedicated DSP microprocessor-like device. Within DSP the same signal processing operations, multiplication, convolution, Fourier transform, modulation, filtering, etc., may be performed. A complete DSP-based system contains the elements shown in Figure 2.37. We mentioned the need for an ADC earlier and here we have its counterpart, a digital to analogue convertor (DAC) to convert digitally processed signals back into analogue form. Both filters are low pass. At the input a filter cuts off any spurious frequencies above the highest frequency of the information signal. This prevents any aliasing when sampling by means of an ADC. An output filter ensures only the desired output spectrum is presented and cuts off any higher-order frequency components, perhaps introduced by the sampling process as illustrated in Figure 2.33 earlier.

There are numerous advantages in processing signals digitally. DSP is very stable and reliable, not subject to drift and variation encountered in some analogue circuits. It also opens up the possibility of performing some operations that are impossible with analogue processing. For instance $g(n)$ may be readily stored in computer memory and therefore does not necessarily need to be processed in real time.

It is interesting to compare the digital representation of a sine wave with its analogue counterpart. Unless the number of samples N per cycle are an exact multiple of the original waves period T, then the digital representation will not be repetitive. This is not a disadvantage as such, merely an observation, for it can be confusing when examining number sequences. Providing the sampling rate is adequate, $g(n)$ nevertheless contains all necessary information and a pure sine wave may still be recovered.

Figure 2.38 illustrates how a sine wave, of period t_1, may be sampled to produce digital samples. Note that in practice ac signals are usually dc shifted to produce unipolar signals in order to match the unipolar characteristics of most ADC circuits. This figure clearly indicates the point above where the sampling rate is not an exact multiple of the frequency of the sine wave. As a result samples from 0 to t_1 are not identical to those produced over the second cycle t_1 to $2t_2$, e.g. $g(1)$ and $g(7)$ are the first samples in two successive cycles but do not occur at exactly the same relative moment in time and hence have different values.

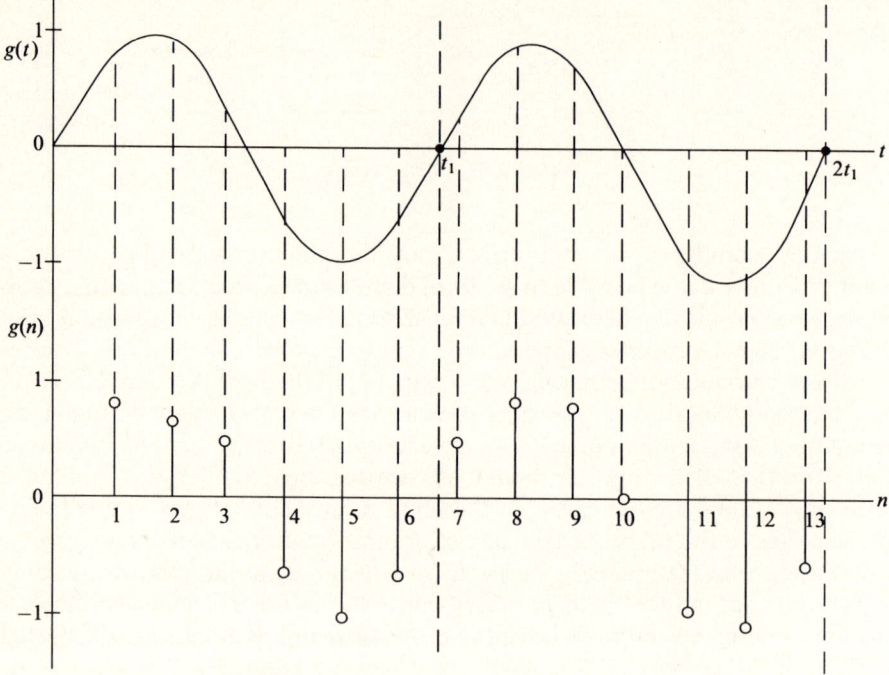

Figure 2.38 Digital representation of a sine wave

2.8 Digital filters

Digital filters are simply filters that are realised by means of DSP techniques. There are two types of digital filters. Firstly consider a **finite impulse response** (FIR) filter shown in Figure 2.39. The first sample, g_0 say, of an incoming number sequence at the input is multiplied by a factor h_0 and appears at the output as k_0. (Note, it is assumed that because there is no signal activity for some while, no sample is present via h_1 through h_3 at this moment.) Each of the paths h_0 through h_3 are known as **taps**. In addition there exists a series of time delays each of duration T which is the period of the sampling process at the receiver. The input pulse g_0 then reaches the multiplier h_1 after one time delay, and by which time g_1 appears at the input. Hence at time $t = 0$, the output k_0 equals the product $g_0 h_0$. At time t equal to T, the output k_1 becomes $g_0 h_1$ plus the new input pulse which appears as $g_1 h_0$. This process continues at interval T and it is suggested that you continue to express k_2, k_3 and so on to establish more fully an FIR filter action over time.

From Section 2.7 we may infer that $k(n)$ is found by convolution of $g(n)$ and $h(n)$.

$$k(n) = g(n) * h(n) \tag{2.78}$$

Input

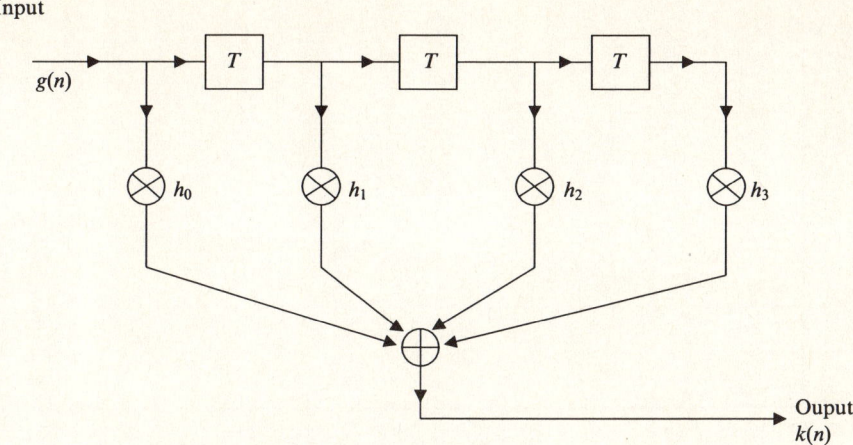

Figure 2.39 FIR filter

$k(n)$ is known as the **convolution sum** $g(n) * h(n)$. The filter output is the sum of the products of $g(n)$ and $h(n)$ for successive n.

In order to test the frequency response of a filter, a signal of constant amplitude with an infinite frequency range is required. This may be achieved by application of an impulse $\delta(n)$. Consider again the filter shown in Figure 2.39. We shall assume that h_1 and h_3 are negative. The output of such a filter is shown in Figure 2.40. Beyond time $3T$ after the original impulse, the filter output remains at zero.

An FIR is so called because, as we have seen, the output response collapses N samples after the original input impulse and where N is simply the number of delay elements, or taps. That is the filter has a **finite impulse response** in time. Figure 2.39 gives us an insight into how an FIR filter may be implemented. A structure is required that multiplies (by each tap weighting), adds and successively stores input impulses. Storage of an impulse as a number, fixed duration delay, addition and multiplication are all commonplace operations which may be performed in hardware, by a digital computer or microprocessor, or by a dedicated DSP system.

Self-assessment 2.8 Sketch the impulse response of the FIR filter shown in Figure 2.41.

Figure 2.42 illustrates an **infinite impulse response** (IIR) filter consisting of three taps. Comparing with the FIR filter seen earlier we may see that all of the taps **feed back** from the output compared with the purely **feed forward** arrangement of an FIR filter.

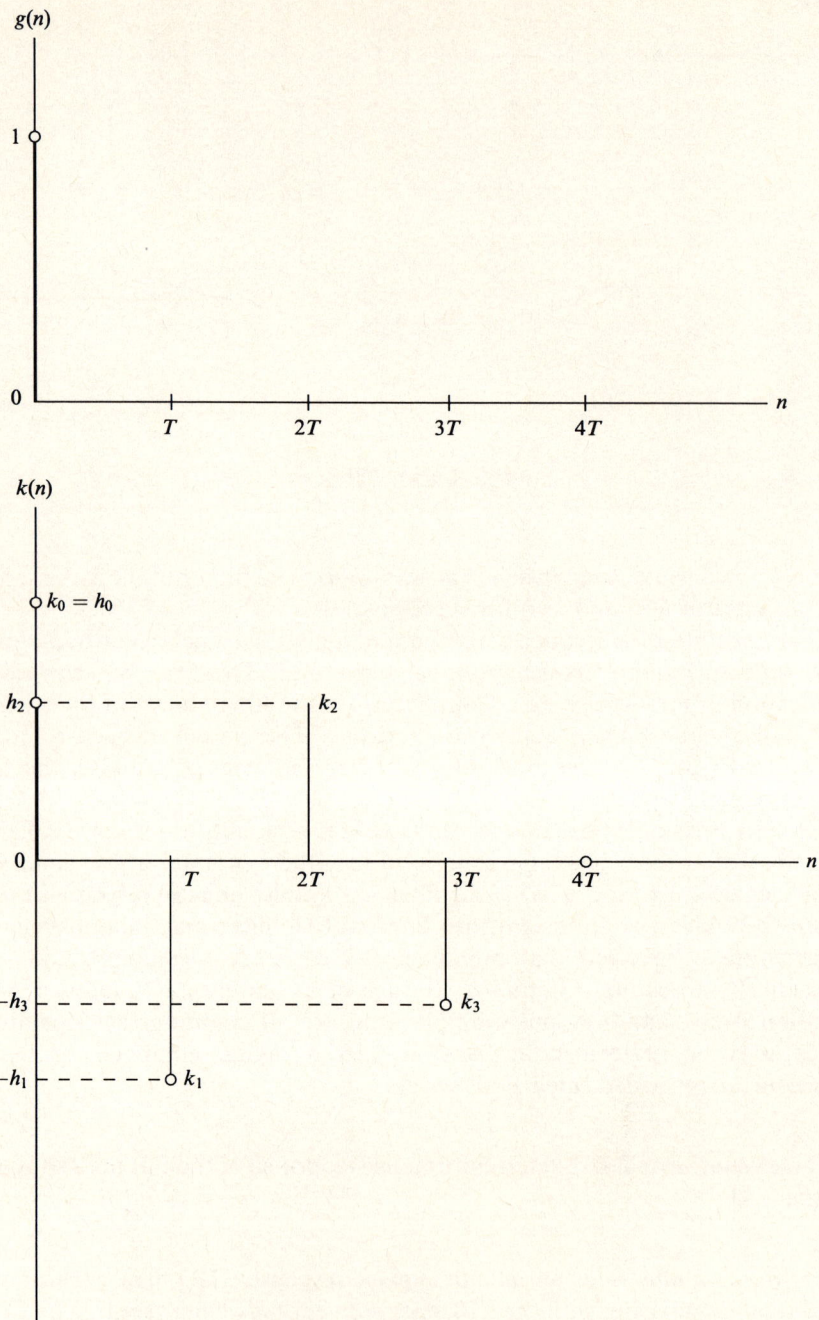

Figure 2.40   FIR filter response

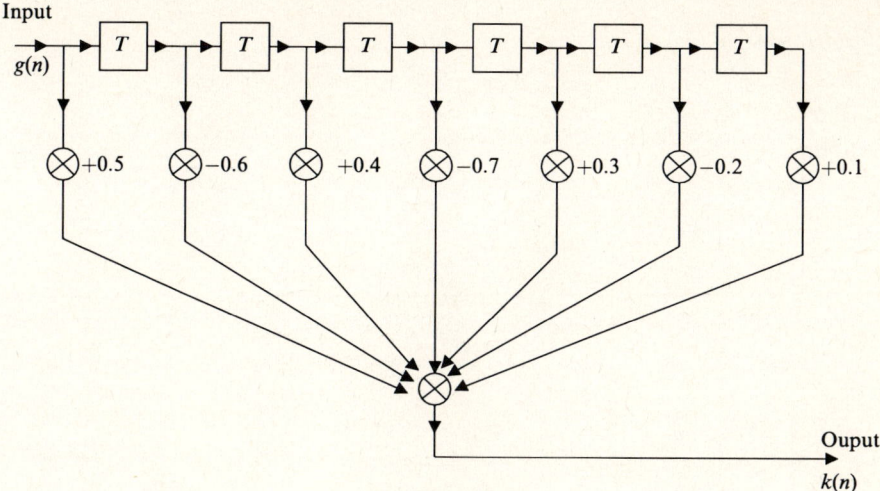

Figure 2.41

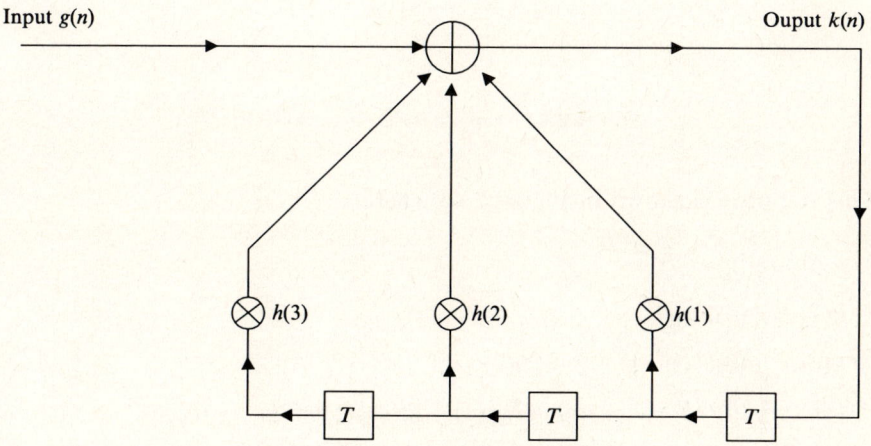

Figure 2.42 IIR filter

The impulse response of an IIR filter is more difficult to determine than that of an FIR. In the figure above we may write an expression for $k(n)$ thus:

$$k(n) = g(0) + [g(0)]h(1) + [g(0)h(1)]h(1) + [g(0)]h(2)$$
$$+ [g(0)h(1)h(1) + g(0)h(2)]h(1) + [g(0)h(1)]h(2) + [g(0)]h(3)$$
$$+ [g(0)h(1)h(1)h(1) + g(0)h(2)h(1) + g(0)h(2)h(1) + g(0)h(3)]h(1)$$
$$+ [g(0)h(1)h(1) + g(0)h(2)]h(2) + [g(0)h(1)]h(3) \text{ etc.} \qquad (2.79)$$

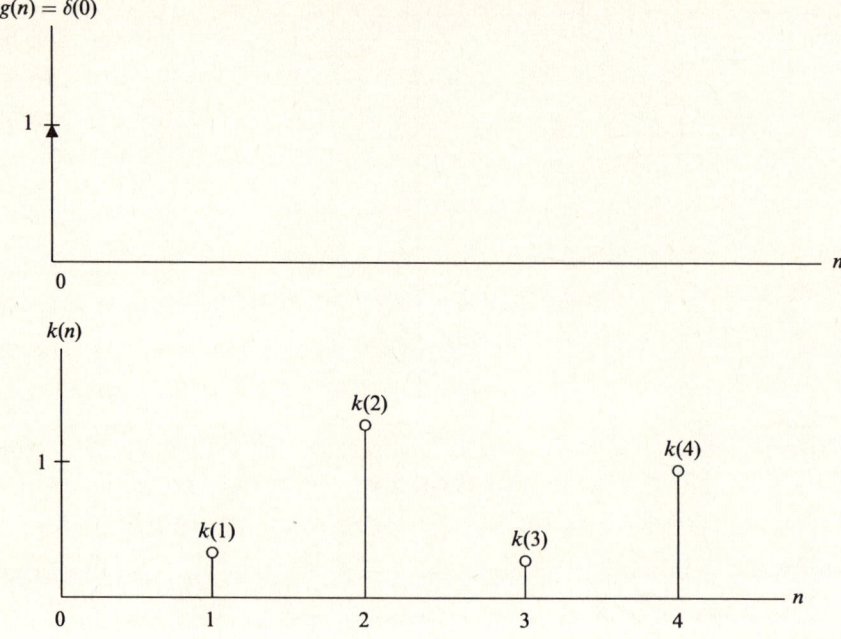

Figure 2.43 Impulse Response of an IIR filter

This response is shown in Figure 2.43 where:

$$k(0) = g(0)$$
$$k(1) = [g(0)]h(1)$$
$$k(2) = [g(0)h(1)]h(1) + [g(0)]h(2)$$
$$k(3) = [g(0)h(1)h(1) + g(0)h(2)]h(1) + [g(0)h(1)]h(2) + [g(0)]h(3)$$
$$k(4) = [g(0)h(1)h(1)h(1) + g(0)h(2)h(1) + g(0)h(2)h(1) + g(0)h(3)]h(1)$$
$$+ [g(0)h(1)h(1) + g(0)h(2)]h(2) + [g(0)h(1)]h(3), \text{ etc.} \qquad (2.80)$$

In order to design a digital filter one can, given the required filter specification in terms of cut-off frequency and roll-off, select a known analogue filter
response, e.g. Chebyshev of given order. Given its transfer function, it is
mathematically possible (e.g. using z-transforms) to determine the equivalent
digital filter transfer function in the form of a number sequence. It is then a
simple matter to sketch the filter structure which may then be realised in
hardware or, more usually, in software using a suitable DSP platform. Such a
design approach is mathematically tedious and prone to error. More usually,

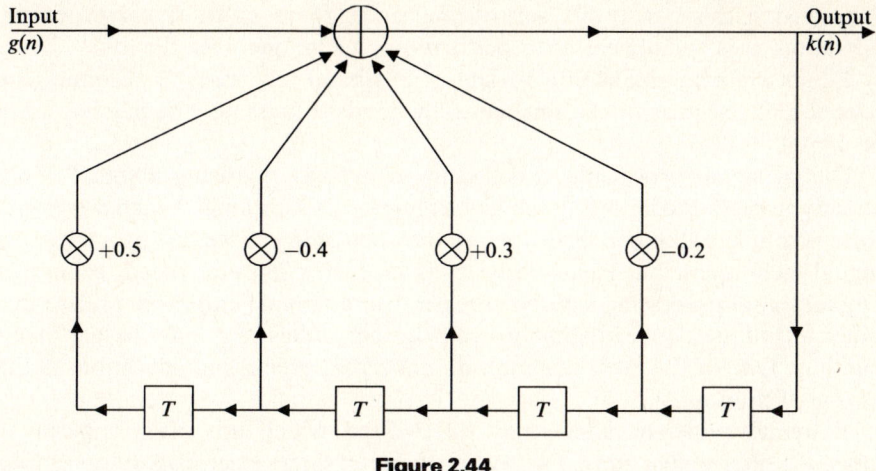

Figure 2.44

DSP is so developed and supported by manufacturers of DSP devices that many filter design packages are available which will directly determine the number of taps and their weightings for a given filter requirement. From a design consideration an IIR filter, for a given rate of cut-off, may be designed with less taps. However, an IIR filter suffers from a poorer phase response compared with that of an FIR filter.

Self-assessment 2.9 Sketch the impulse response of the IIR filter shown in Figure 2.44.

2.9 Summary

The relationship between a signal's time and frequency domain relationship may be found using Fourier techniques. The Fourier series enables the spectrum of a function that is periodic in the time domain to be determined. The Fourier transform may be used to determine the spectrum of a non-periodic signal. The inverse Fourier transform enables the time domain representation of a signal to be determined from its spectrum.

Many signals are of standard form, or may be represented by a combination of standard forms. In such instances the sometimes lengthy calculations required of Fourier techniques may be avoided by the use of well-published results of Fourier series, or transforms. Where a signal is continuous in time, its spectrum is finite, and vice versa. This fact is useful in considering the bandwidth of a signal.

Convolution and multiplication are reverse, or dual, processing operations. Convolution in one domain is equivalent to multiplication in the other.

Recognising this can prove very useful since, for a given signal processing operation, it is usually easier to perform one technique over the other.

Filters are available with a variety of passband responses to accommodate selection or rejection of certain frequency bands. Filters may be passive, active or digital.

Digital signal processing is based upon initially digitising a signal, if not already in digital form, using the key technique of sampling. A signal in digital form is simply a number sequence and may be readily processed by a variety of digital techniques, but particularly lends itself to computer-based techniques. Digital signal processing is often cheaper, more reliable and more precise than other techniques. Indeed some processing techniques may only be performed digitally. One of the most common digital signal processing operations is that of digital filtering.

Extremely efficient filters may be realised which are often superior to analogue filter realisations. Two types of digital filters exist. FIR filters employ feedback taps which lead to an impulse response which collapses after a finite time interval dependent upon the number of taps. Alternatively, IIR filters employ feedback giving rise to an impulse response which persists infinitely in time. Both types of digital filter have their relative merits.

Exercises

2.1   For each of the repetitive waveforms shown in Figure 2.45:

(a)   Determine if any symmetry is exhibited.
(b)   Determine an expression for the frequency spectrum of each waveform. Make use of any symmetry to simplify your calculations. Where possible, confirm your results using Fourier series tables.

2.2   With reference to the function $g(t)$ shown in Figure 2.46, determine:

(a) $g(-t)$
(b) $g(t/4)$
(c) $g(3-t)$
(d) $g(t/4+1)$
(e) $g\dfrac{(t+1)}{4}$

2.3   Determine the Fourier transform of the signal shown in Figure 2.47 by:

(a)   means of first principles, based upon Eqn (2.25);
(b)   using the Fourier transform tables, based upon the fact that $g(t)$ may be regarded as a pair of rectangular functions, each of which is time-shifted.

(i)

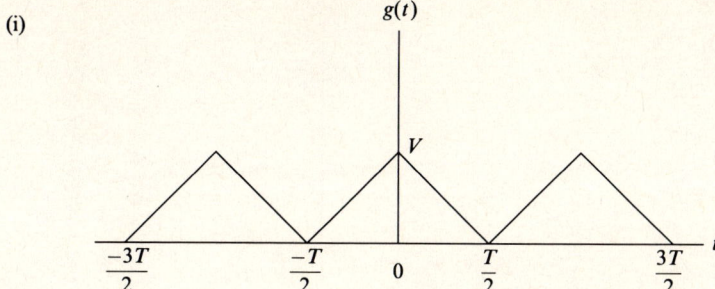

(ii)

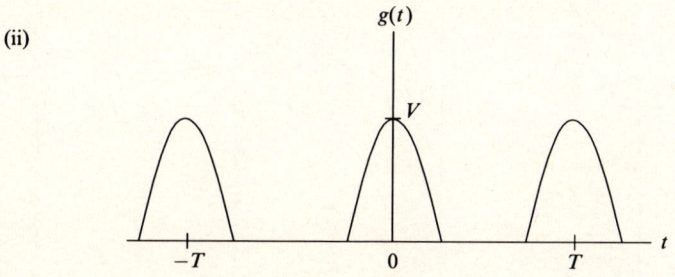

(iii)

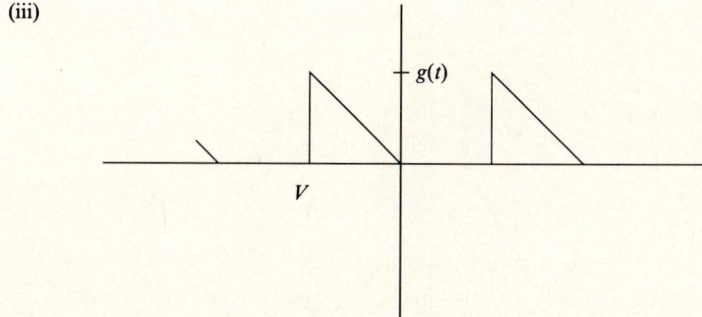

Figure 2.45

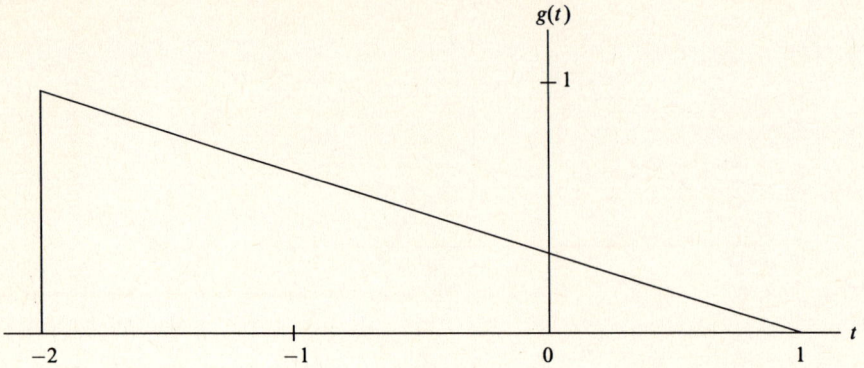

Figure 2.46

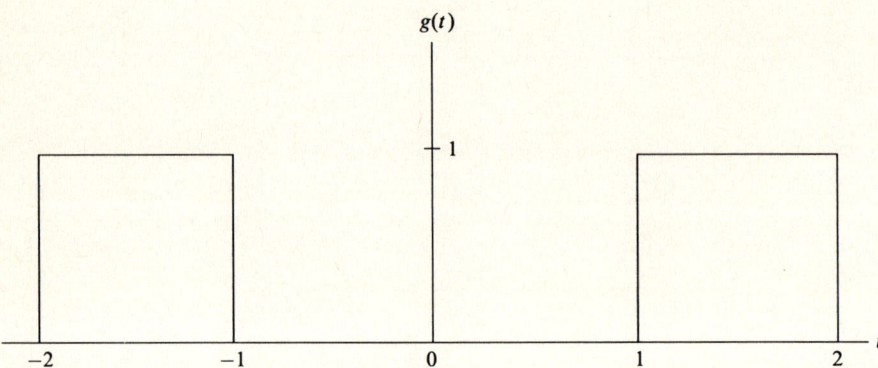

Figure 2.47

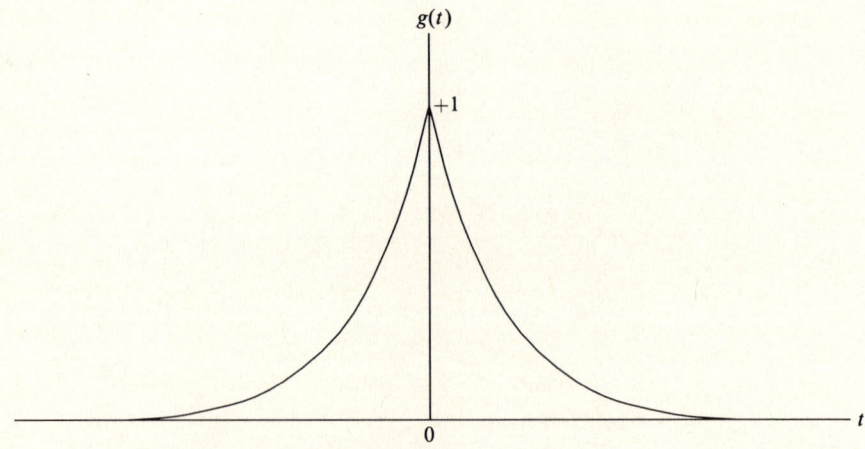

Figure 2.48

(a)

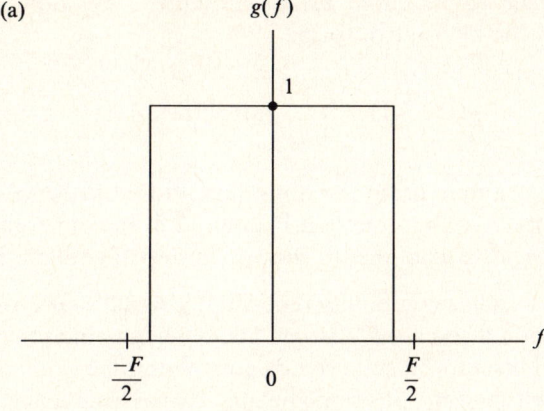

(b)

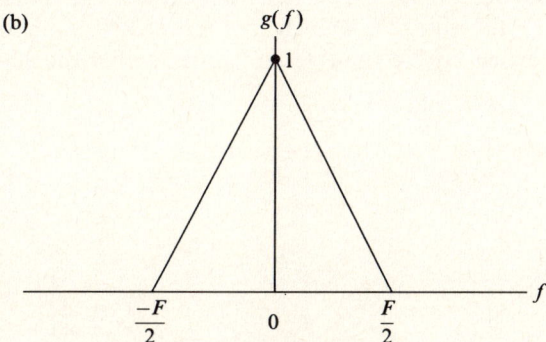

Figure 2.49

2.4 The waveform shown in Figure 2.48 is defined:

$$g(t) = e^{-t}, \qquad t < 0$$

$$= e^{t}, \qquad t - 0$$

Determine the Fourier transform of $g(t)$ by means of first principles. Confirm your answer by use of Fourier transform tables.

2.5 For of each of the functions shown in Figure 2.49, determine the time domain response:

(a) by first principles;
(b) by making use, where possible, of Fourier transform tables.

Make a sketch of your responses.

2.6 With reference to Example 2.8 and Figure 2.27, Eqn (2.59) may be expressed in the alternative form:

$$g(t) * h(t) = \int_{-\infty}^{\infty} g(t) \cdot h(t - u)\, du \tag{2.81}$$

Rework the example using the above equation and hence show that the result is identical to that found in Example 2.8. In your solution, include a series of illustrative examples to show each step of convolution pictorially.

2.7 Many radio systems commonly reposition a signal's spectrum such that it is centred upon a so-called 'carrier' frequency at which the system is said to operate. This may be achieved by convolving the original signal with a pair of delta functions positioned at the carrier frequency.

A particular signal and the carrier frequency of the radio channel to be used to carry the signal is shown in Figure 2.50. Convolve these two spectra and hence, with a suitable sketch, show that the above statements are correct.

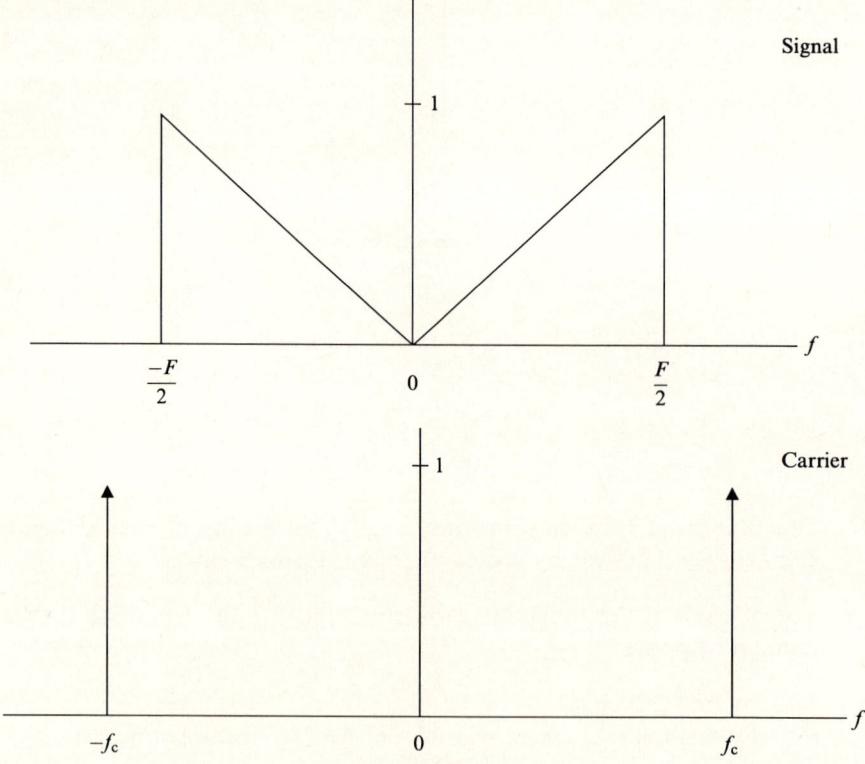

Figure 2.50

2.8   Explain the effect of aliasing and its interdependence upon both frequency content of a signal and sampling rate.

2.9   With reference to digital filters, discuss what is meant by:

  (a)   FIR;
  (b)   IIR;
  (c)   feed forward;
  (d)   feedback.

Bibliography

Bracewell, R.N., *The Fourier Transform and its Applications*, 2nd edn, McGraw-Hill, 1978. ISBN 07-066196-0.

Carlson, A.B., *Communication Systems*, 3rd edn, McGraw-Hill, 1986. ISBN 0-07-100560-9.

James, G, *Advanced Modern Engineering Mathematics*, Addison Wesley, 1993. ISBN 0-201-56519-6.

Lynn, P.A. and Fuerst, W., *Introductory Digital Signal Processing with Computer Applications*, John Wiley, 1994. ISBN 0-471-94374-6.

Stremler, F.G., *Introduction to Communications Systems*, 3rd edn, Addison-Wesley, 1990. ISBN 0-201-51651-0.

Stroud, K.A., *Further Engineering Mathematics*, Macmillan, 1990. ISBN 0-333-52610-4.

CHAPTER 3
Noise

Aims and objectives

A definition for the term noise is established. A number of causes of noise are identified and modelled to enable noise to be quantified in electrical circuits. Thermal noise is a major source of noise and its properties and relationship to temperature are explained. A model is developed to represent noise introduced by an electrical network in terms of effective noise temperature of a source resistor. Noise peculiar to radio systems is discussed. Noise received by an antenna is represented as an equivalent noise temperature and typical values are presented. The concept of effective noise temperature is extended to represent the noisiness of a receiver by expressing it in terms of a noise temperature at a receiver input. A signal at any point within a system is always accompanied by noise. The concept of signal to noise ratio is introduced and it is shown that a certain amount of noise may be tolerated in a communication system. An alternative approach to signal to noise ratio is established to represent the noise contributed within a system by a network by defining the noise figure of a network. The noise figure enables the overall noise performance and the signal to noise ratio to be readily calculated for a system comprising a number of networks in cascade.

3.1 Introduction to noise

Electrical **noise** may be defined as any unwanted energy that accompanies a signal in a communication system. A signal at any point within a system is always accompanied by noise which is in general due to the cumulative effects of a number of similar and dissimilar causes, many of which are described in this chapter. Initial thoughts may be that the presence of noise in a system might prove to be disastrous. While this may be true on occasion, most systems perform satisfactorily with a signal that is accompanied by a limited level of noise.

One example of noise is **crosstalk** where, for example, during a telephone system the conversation in one telephone connection is audible in another. Another form of noise is **co-channel interference** occasionally experienced in some television systems where under certain abnormal atmospheric conditions transmissions may propagate over a greater distance than normal and cause interference with local transmissions on a similar frequency.

Noise in communication systems falls into two types. Some is **manmade** (or **artificial**) and could be eliminated through better design, perhaps by suppression at source, or screening of offending sources or sensitive circuit elements. Artificial noise arises from many sources such as electrical machinery and switches and certain types of lamps. For example vehicle ignition systems may, if not properly suppressed, interfere with domestic radio and television broadcast reception.

Many other forms of noise are unavoidable in that they occur **naturally**. We are all familiar with crackles on our radios at home due to lightning discharges. This is an example of **atmospheric noise**. Another source of noise is radiation from space known as **cosmic noise** (or space noise) which is emitted by stars as a result of energy conversion. In addition, as we shall see in the next section, many electrical and electronic components naturally introduce noise into a system.

3.2 Noise in electronic components

Electric circuits are in general formed from combinations of passive and active components. An output signal from a circuit may be regarded as arising from either a voltage or a current source. Many circuits also produce an extraneous noise voltage, or current, at the output in addition to that of the signal. Some noise may be introduced by one or more circuit components by means of the mechanisms described below.

3.2.1 Thermal noise

Within a conductor or resistor free electrons are produced because of thermal agitation. These electrons move randomly resulting in a rate of arrival at each end that also varies in a random manner. This in turn gives rise to a randomly varying potential difference across the ends of the conductor or resistor, although it has a mean value of zero. Such a noise source is termed **thermal noise** because its energy increases with temperature.

Figure 3.1 shows a simple model of a resistor and an associated noise voltage e_n. The noise voltage varies in time with a Gaussian probability distribution function and, as mentioned, mean value of zero. This means that it is alternating, and it is uniformly distributed spectrally from 0 to about 10^{13} Hz which is frequencies far in excess of those used in communication circuits. If a short-circuit is placed across the ends of the resistor, the noise power developed within resistor R is given by:

$$\text{Noise power} = 4kTB \qquad (3.1)$$

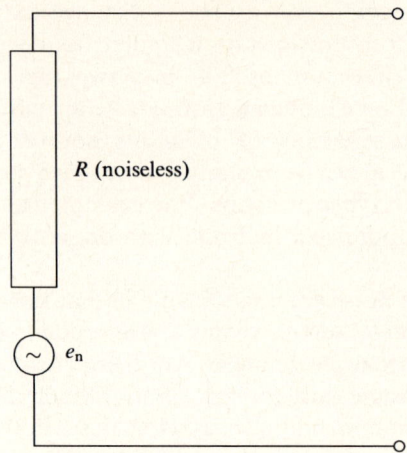

Figure 3.1   Thermal noise model

where: k is Boltzmann's constant and equals 1.38×10^{-23} W/kHz;
T is absolute temperature in kelvin, K,
B is bandwidth under consideration, e.g. of the measuring instrument
or system (note that to conform with many other works, bandwidth in
this chapter will be denoted by the symbol B rather than W used
elsewhere).

The noise voltage e_n is given by:

$$\frac{e_n^2}{R} = 4kTB \tag{3.2}$$

$$\therefore \quad e_n = \sqrt{4kTBR} \tag{3.3}$$

Equation (3.3) indicates that the voltage developed by a resistor because of
thermal noise is a function of temperature, bandwidth and resistor value.

What is important in practical systems is the noise delivered into a load.
Usually, in a communication system, source and load are matched in order to
achieve maximum power transfer. The **available noise power** is the maximum
power that may be developed in a load resistance. Consider a noisy source
connected to a load resistor, R_L, (noiseless) as shown in Figure 3.2. The
available noise power developed in R_L is:

$$P_L = \frac{\overline{e_n^2} \cdot R_L}{(R_s + R_L)^2} \tag{3.4}$$

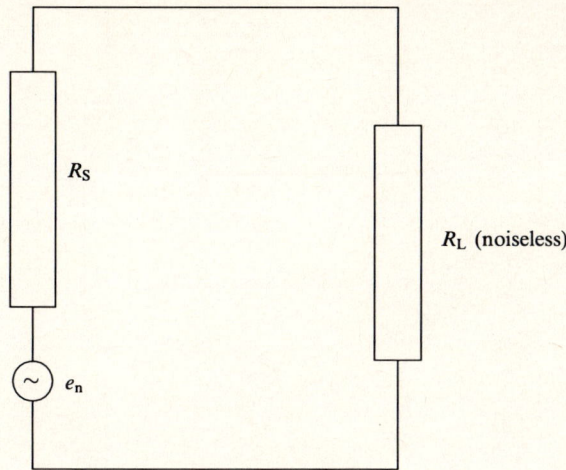

Figure 3.2 Available noise power

where e_n is the noise generated in the source resistor, R_s, at temperature T_s. With matched conditions where $R_s = R_L$, P_L becomes:

$$P_L = \frac{\overline{e_n^2}}{4R_L} \qquad (3.5)$$

From Eqn (3.3) we may substitute e_n into Eqn (3.5):

$$P_L = \frac{4kT_sR_sB}{4R_L} \qquad (3.6)$$

$$= kT_sB \qquad (3.7)$$

Hence, under matched conditions, the available noise power in a given bandwidth B delivered into a resistive load is readily found by knowledge of the temperature of the source resistance. In practice the 'load' may be the input resistance of an intermediate sub-system.

EXAMPLE 3.1
Determine an equivalent noise circuit to represent two resistors R_1 and R_2 connected in series at temperatures T_1 and T_2 respectively.

SOLUTION
When connected in series we may draw an equivalent circuit as shown in Figure 3.3. Since both e_1 and e_2 are rms (root mean square) values, we

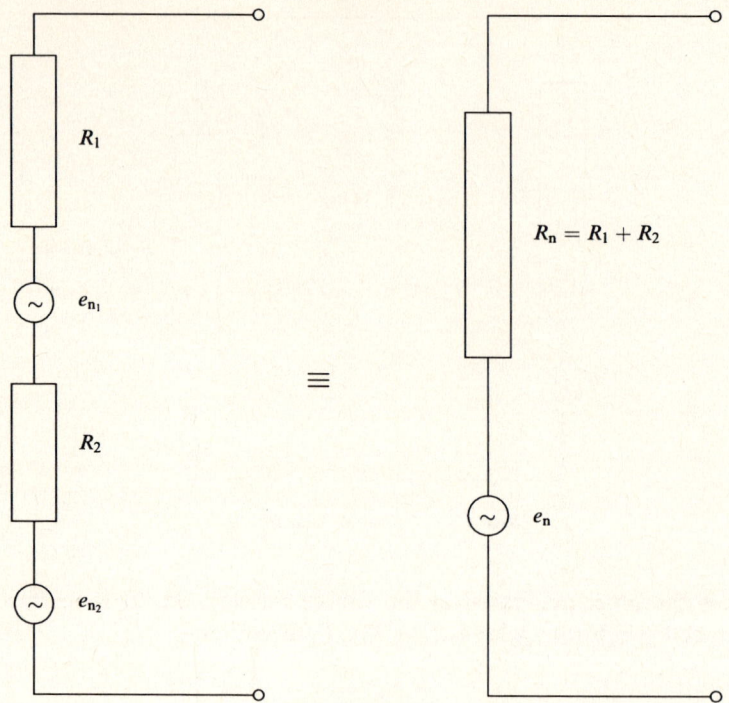

Figure 3.3

may add their mean squared values:

$$\overline{e_n^2} = \overline{e_1^2} + \overline{e_2^2} \tag{3.8}$$

$$= 4kBT_1R_1 + 4kBT_2R_2 \tag{3.9}$$

Hence,

$$e_n = \sqrt{[4kB(T_1R_1 + T_2R_2)]} \tag{3.10}$$

In practice the two resistors in series are invariably at the same temperature which we may represent by T. Therefore we may write:

$$e_n = \sqrt{[4kTB(R_1 + R_2)]} \tag{3.11}$$

Hence we see that an arrangement of two resistors in series at the same temperature behave as a single resistor with resistance equal to that of the series equivalent operating at the same temperature.

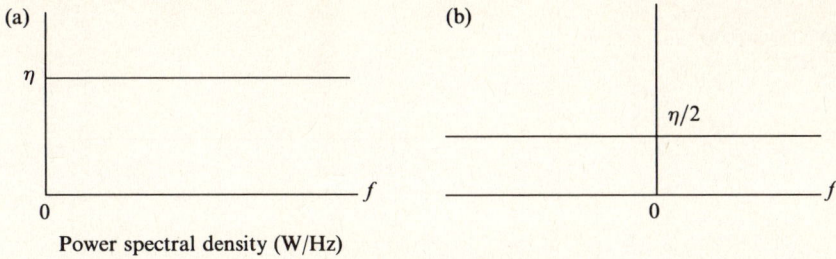

Power spectral density (W/Hz)

Figure 3.4   Thermal noise spectrum: (a) single sided; (b) double sided

The power spectral density (PSD) of thermal noise is indicated in Figure 3.4 where, in the case of the single sided representation, η is the normalised value of PSD in watt per hertz. Hence the power in a given bandwidth B is simply ηB watts. By analogy with that of white light, thermal noise is therefore also known as **white noise**. It is also called **Johnson** noise after the name of an early experimenter with noise.

Example 3.1 illustrated that the total mean squared noise voltage across two, or more, resistors in series equals the sum of their individual mean squared voltages. This additive property, coupled with a white noise spectrum, gives rise to another name for thermal noise, namely **additive white Gaussian noise** (AWGN).

In practice systems do not operate over the full range of frequencies over which thermal noise may be experienced. Suppose, under matched conditions, a noise source is passed to a load via an intermediate network which is a bandpass filter, as shown in Figure 3.5. If the noise contribution of the filter is ignored, the output noise spectrum is the product of the input noise spectrum and the filter response. The actual noise power at the output equals the area under the output PSD curve (which may be found by integration). This power is equivalent to the area under an 'ideal' curve with a rectangular PSD of bandwidth B, as shown, of same area as that of the actual output PSD curve. The value of B is known as the **equivalent noise bandwidth**.

3.2.2 Shot noise

Shot noise is the next most significant form of electrical noise after thermal noise. Where a mean, or dc, current is flowing through an active device (semiconductor or thermionic), there is a random variation in current superimposed upon the dc value. This is due to variation in the arrival time of charge carriers at the output electrode which gives rise to a randomly varying noise current, in addition to the mean value.

Shot noise, unlike thermal noise, only occurs in active devices. It does, however, for all practical purposes, exhibit similar properties to those of thermal

Input noise power (W/H_z)

$\frac{n}{2}$

0

f

Bandpass filter transfer function

1

0

f_0

f

Equivalent noise bandwidth, B

$\frac{\eta}{2}$

Load noise power (W/H_z)

0

f_0

f

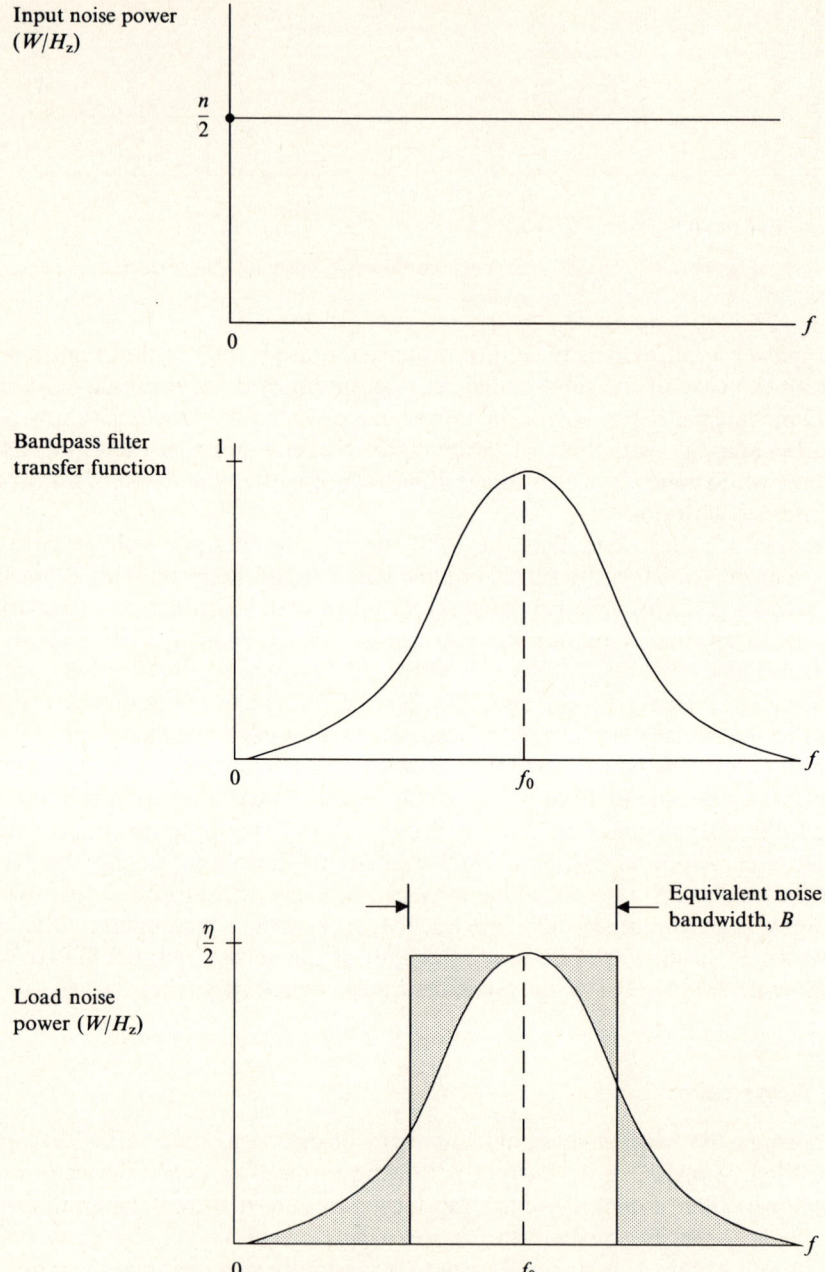

Figure 3.5 Equivalent noise bandwidth

noise and contributes a noise power proportional to bandwidth. It derives its name from the way that when amplified and the signal directed to a loudspeaker it sounds like a series of lead shots falling on a metal sheet.

An equivalent circuit to represent shot noise for a semiconductor diode in conduction is shown in Figure 3.6. For frequencies up into the Gigahertz range, the rms value of shot noise current I_n is given by the expression:

$$I_n = \sqrt{(2qI_{dc}B)} \tag{3.12}$$

where: q is the charge on an electron $= 1.6 \times 10^{-19}$ C,
         I_{dc} is the dc, or bias, current, and
         B is the bandwidth.

Note how shot noise is proportional to the square of the mean current flowing through a device. The dynamic junction resistance r_d is given by:

$$r_d = \frac{kT}{qI_{dc}} \tag{3.13}$$

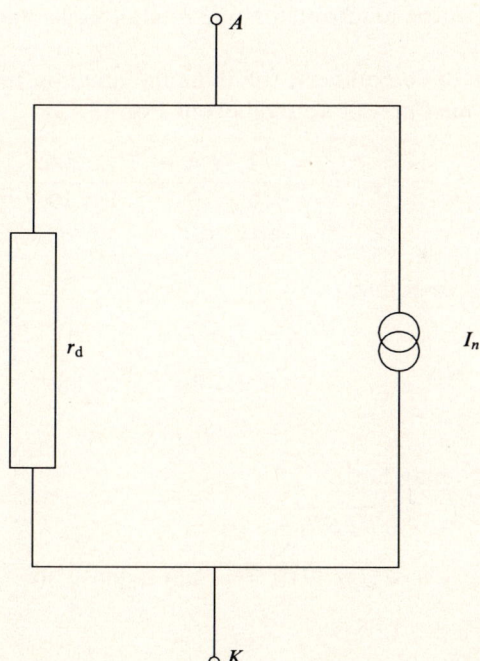

Figure 3.6   Shot noise model (diode)

EXAMPLE 3.2

A semiconductor signal diode passes a mean value of current equal to 0.1 mA. Calculate the noise current produced when operating over a bandwidth of 5 MHz.

SOLUTION

$$I_n = \sqrt{(2qI_{dc}B)} \tag{3.14}$$

$$= \sqrt{(2 \times 1.6 \times 10^{-19} \times 0.1 \times 10^{-3} \times 5 \times 10^{6})} \tag{3.15}$$

$$= 12.7 \, \text{nA} \tag{3.16}$$

The noise voltage produced due to shot noise may also be determined for a given electrical circuit. Consider the following example.

EXAMPLE 3.3

If a diode, operating at 27°C, passing a mean value of current of 1 mA is connected in series with a load resistance of 2 kΩ, determine the noise voltage developed in the load. The bandwidth may be assumed to be 200 kHz.

SOLUTION

The voltage generated in the load due to shot noise may be found using the equivalent circuit shown in Figure 3.7 where v_s and v_t are the voltages that occur in the load resistor due to shot noise and thermal noise, respectively.

In order to determine v_s the dynamic junction resistance r_j, which equals r_d, must first be found. From Eqn (3.13):

$$r_d = \frac{1.38 \times 10^{-23} \times 300}{1.6 \times 10^{-19} \times 1 \times 10^{-3}} \tag{3.17}$$

$$= 25.88 \, \Omega \tag{3.18}$$

v_s may now be found:

$$v_s = I_{dc}r_d \tag{3.19}$$

$$= 1 \times 10^{-3} \times 25.88 \tag{3.20}$$

$$= 25.88 \, \text{mV} \tag{3.21}$$

Similarly we may find v_t:

$$v_t = \sqrt{kTBR_L} \tag{3.22}$$

$$= \sqrt{1.38 \times 10^{-23} \times 300 \times 200 \times 10^3 \times 2 \times 10^3} \tag{3.23}$$

$$= 1.287 \, \mu V \tag{3.24}$$

Now the voltages generated independently by v_s and v_t in R_L is simply found by voltage division in the two series resistors.

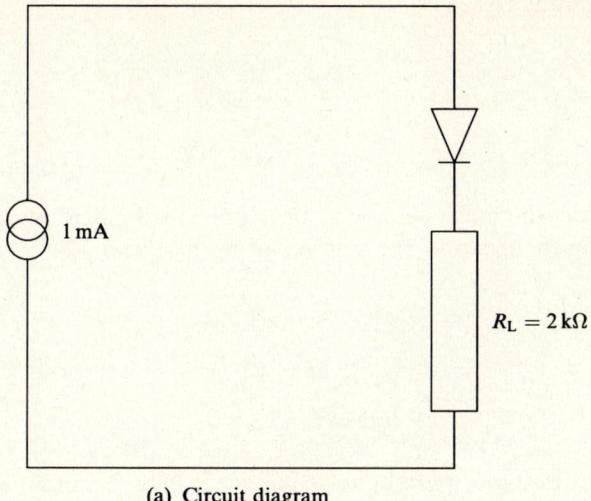

(a) Circuit diagram

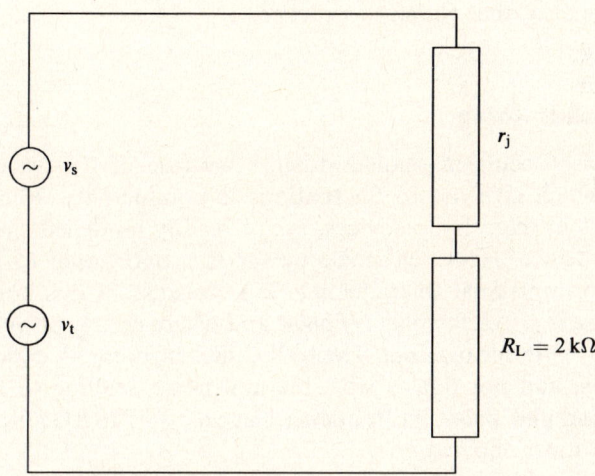

(b) Equivalent noise circuit

Figure 3.7

The voltage across R_L due to v_s, which we shall indicate as v'_s, may be found:

$$v'_s = \frac{25.88 \, \text{mV} \times 2 \, \text{k}\Omega}{25.88 \, \Omega + 2 \, \text{k}\Omega} \tag{3.25}$$

$$= 25.55 \, \text{mV} \tag{3.26}$$

And similarly,

$$v'_t = \frac{1.287\,\mu V \times 2\,k\Omega}{25.88\,\Omega + 2\,k\Omega} \tag{3.27}$$

$$= 1.271\,mV \tag{3.28}$$

Now the noise voltage v_n developed in the load resistor is found by taking the mean of the squares of both v'_s and v'_t:

$$v_n = \sqrt{\overline{(v'_s)^2} + \overline{(v'_t)^2}} \tag{3.29}$$

$$= \sqrt{[(25.55 \times 10^{-3})^2 + (1.271 \times 10^{-6})^2]} \tag{3.30}$$

$$= 2.55\,mV \tag{3.31}$$

In this example thermal noise voltage is in excess of an order greater than that of shot noise, the total noise voltage being virtually due to thermal noise alone. Such relative values for thermal and shot noise are not atypical of that found in practice.

3.2.3 Flicker noise

Flicker noise occurs in semiconductors because of defects in the crystalline material which give rise to fluctuations in conductivity. Flicker noise, which also occurs in thermionic devices, is not readily modelled since it varies from device to device. However, noise power is proportional to bias current and inversely proportional to frequency. It is because of this latter property that flicker noise is also known as **1/f noise** and where clearly power diminishes with frequency. This means that spectrally, flicker noise is concentrated at low frequencies, and not flat as with thermal noise, leading to flicker noise also being termed **pink noise**. At frequencies in excess of 10 kHz flicker noise may be ignored in most applications.

3.3 Effective noise temperature

The available noise power developed in a load by a source may be found by means of an equivalent circuit, such as a Thévenin equivalent circuit, to represent the circuit. It is then a straightforward matter to determine the available noise power developed in a load. Where instead a source is interconnected to a load via an intermediate network, it is less obvious what the available noise power will be in a load, particularly, as is usual in practice, the network itself introduces noise. Such a system may be represented by Figure 3.8.

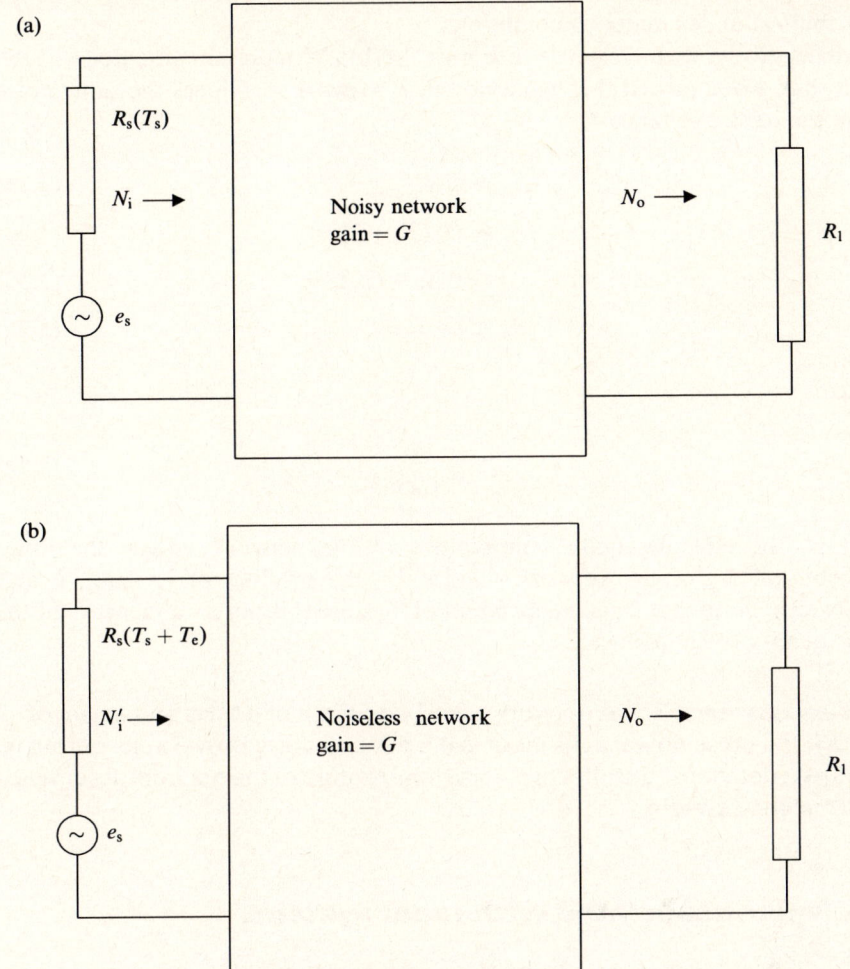

Figure 3.8 Effective noise temperature: (a) noisy and (b) noiseless network

The available noise power in the load due to noise contributions from the source and network may be represented by a noise temperature associated with the source resistance R_s which is the sum of T_s, the source resistance temperature, and T_e the **effective noise temperature** of the network. The effective noise temperature is that temperature associated with R_s which develops the same noise power in the load as the contribution purely due to the network. Under matched conditions:

$$N_o = GN_i + N_n \tag{3.32}$$

where N_n is the noise developed in the load purely due to the network contribution under matched conditions.

Alternatively, with reference to Figure 3.8(b), N'_i represents the noise at the input that, when passed through a noiseless network, produces the same noise N_o in the load resistance:

$$N_o = GN'_i \qquad (3.33)$$

$$= Gk(T_s + T_e)B \qquad (3.34)$$

$$= GN_i + GkT_eB \qquad (3.35)$$

$$\therefore \quad N_n \equiv GkT_eB \qquad (3.36)$$

Hence:

$$T_e = \frac{N_n}{GkB} \qquad (3.37)$$

That is, the effective noise temperature of the network equals the noise contribution N_n of the network divided by the product of its gain, k and bandwidth. Note that T_e is independent of T_s since it is purely a measure of the noise quality of the network.

Self-assessment 3.1 A network has a bandwidth of 6 MHz and a gain of 2000. If the noise power at its input is 0.5 pW and noise power at its output is 1008 pW, determine the effective noise temperature of the network assuming a 50 Ω matched system.

3.4 Noise associated with radio systems

Noise at radio frequencies may occur naturally. For instance electrostatic discharge, commonly in the form of lightning within the earth's atmosphere, produces **atmospheric noise**, or **static**, over a range of frequency not exceeding about 20 MHz. Where intermittent discharges occur they appear as impulse noise in a receiver. This type of noise may be received at great distance and it is therefore possible for a receiver to experience a large number of discharges per unit time. On occasion discharges may occur so frequently that the noise received appears continuous, rather than impulsive. Another atmospheric effect is **quantum noise** which becomes increasingly significant above 10 GHz. Quantum noise arises from changes in the atomic and molecular states of particles within the atmosphere. In addition the earth produces thermal radiation known as **earth noise**. At frequencies above 200 MHz earth noise becomes significant, increasing in intensity with increasing frequency.

Apart from effects associated with the earth and its atmosphere, naturally occurring noise also occurs from space and is termed **cosmic noise**. The sun radiates appreciable noise, **solar noise**, between 100 MHz and about 10 GHz. Other stars within the galaxy also emit radio frequency (RF) radiation but owing to the vast distance of stars from the earth, such noise is only significant up to 1 GHz.

Abrupt changes in the flow of current, which may occur regularly or as occasional pulses, have a very large spectral content. (This is to be expected by an appreciation of the Fourier transform dealt with earlier in Section 2.3.) These changes may, where sufficiently large electromagnetic fields at radio frequency occur, cause interference with nearby radio receiving equipment. Such **artificial interference** is another example of noise and is especially caused by the switching off of power supplies to electrical equipment containing inductive elements. Suppression techniques should be employed wherever there is a potential for radio frequency interference (RFI). For instance it is good practice to place a suitable capacitor across the contact breaker in a vehicle ignition systems to limit the RF emission caused by arcing when the contact opens. Artificial interference ranges in frequency up to some hundreds of MHz.

In addition to noise produced within electrical circuits of transmitters and receivers, radio systems are especially prone to noise introduced via a receiving antenna due to atmospheric, cosmic and artificial noise. Noise introduced through an antenna is sometimes known as **external noise**, whereas noise that occurs naturally within the receiver is called **internal noise**. The presence of external noise in radio systems makes them potentially more vulnerable to noise than systems that make use of transmission lines for connection of transmitter and receiver.

The combined radiation effects from both atmosphere and cosmos can be regarded as **sky noise** which, with the exception of an antenna pointing straight at the sun, is relatively small compared with internal noise. Sky noise, although ranging over a wide band of frequency, may be regarded as spectrally flat over the relatively narrow band of a particular system. Sky noise is a minimum at microwave frequencies of the order of a few GHz which is why many early satellite systems operated at such frequencies, for example 6/4 GHz, to make use of the low-noise 'window'. Beyond 10 GHz sky noise is characterised by atmospheric noise, increasing with frequency.

Sky noise introduced into an antenna may be represented as an effective noise temperature T_{ae} with a bandwidth B associated with an antenna's resistance. Such a 'temperature' is an extremely useful concept for analysis of the noise performance of radio systems. T_{ae} is not normally the actual temperature of the antenna but rather is found by measurement of the received radiation and equated to a noise temperature associated with the antenna's resistance.

In practice manufacturers of antennas may specify T_{ae} at the antenna terminal as part of their specification to include sky noise and antenna noise. Alternatively the effective noise temperature may, in addition to the antenna, also include the noise contribution due to a pre-amplifier and feeder (the conductor arrangement connecting antenna output terminals to the receiving equipment) to which an antenna is often immediately connected. In this case the effective noise temperature is that found at the antenna input to a receiver which we may denote as T_{ai}. This latter term is very useful because a receiver input is a convenient reference point to represent its noise contribution at the output as an effective noise temperature T_e. The overall value of noise temperature of the system may then be found and is simply the sum of the individual temperatures, that is T_{ai} and T_e of the receiver.

A typical satellite communication system has effective noise temperatures of 20 K at the receiver and 20 K at the antenna compared with a high frequency (HF) radio system with figures of 290 K and 2000 K, respectively. Clearly the noise introduced within a satellite receiver is considerably lower. This is essentially because the received signal to noise ratio is relatively low owing to the excessive path length and associated attenuation that characterise satellite communication systems.

Now consider the effective noise temperature of a passive network such as an antenna feeder. A passive network may be represented as shown in Figure 3.9, where it is assumed to be fully matched and hence source, input, output and load resistance are equal and here represented as R.

It is also reasonable to assume that all the components of the network are at the same physical temperature T. The noise power delivered to the load is simply:

$$N_o = kTB \tag{3.38}$$

The output noise power N_o depends upon the noise power at the input N_i, the gain of the network and the noise contribution of the network N_n. Output noise power may therefore be expressed:

$$N_o = GN_i + N_n \tag{3.39}$$

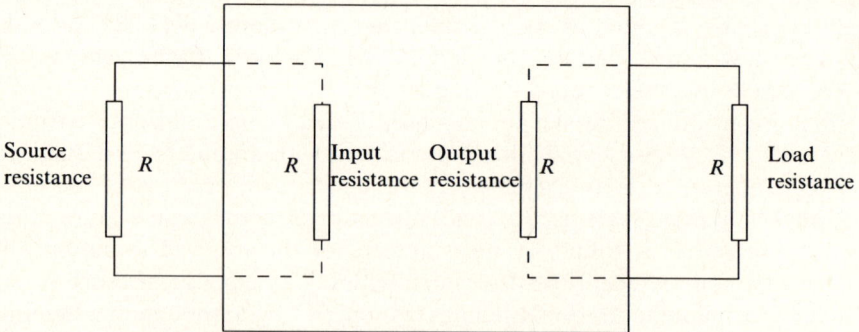

Figure 3.9   Effective temperature of a passive network

Under matched conditions the available noise power at the input equals the noise power developed in the source resistance. Hence:

$$N_i = kTB \tag{3.40}$$

We may in Eqn (3.39) make substitutions for both N_o and N_i using Eqns (3.38) and (3.40), respectively. Therefore:

$$kTB = GkTB + N_n \tag{3.41}$$

Hence:

$$N_n = (1 - G)kTB \tag{3.42}$$

From Eqn (3.37) we may write that the effective noise temperature of the network T_e is given by:

$$T_e = \frac{N_n}{GkB} \tag{3.43}$$

In Eqn (3.43) we may make a substitution for N_n by means of Eqn (3.42). Hence:

$$T_e = \left(\frac{1 - G}{G}\right)T \tag{3.44}$$

In the case of a feeder where gain is less than 1, Eqn (3.44) may be rearranged into more convenient form using loss L, which is the inverse of gain:

$$T_e = (L - 1)T \tag{3.45}$$

Self-assessment 3.2 A passive network has a load resistance of 75 Ω and gain of 0.25. The available noise bandwidth is 8 MHz. If the network is matched and the ambient temperature throughout is 27°C, calculate:

(a) the effective noise temperature of the network;
(b) the noise power in the load;
(c) the noise contribution in the load purely introduced by the network.

3.5 Signal to noise ratio

In most systems it is not the absolute level of noise power that determines what is acceptable to a user, but rather its relative level compared with that of the signal. The ratio of signal power to noise power, or signal to noise ratio (S/N), at a systems output is therefore of paramount importance where the S/N is defined as:

$$\frac{S}{N} = \frac{\text{Signal power}}{\text{Noise power}} \qquad (3.46)$$

Often in practice a system may operate over a large dynamic range, e.g. two or three orders, which requires handling very small and very large values of the S/N ratio. Alternatively S/N is often expressed more conveniently in logarithmic form using the unit **decibel** (dB) where

$$\frac{S}{N} \, dB = 10 \log_{10} \frac{S}{N} \qquad (3.47)$$

For a signal to be usable at the output of a receiver the signal power must exceed that of noise by a acceptable margin. A useful analogy to illustrate this point is that of two people holding a conversation some distance from a pneumatic drill in use for road repair. Speech represents 'signal', and sound from the drill, 'noise'. Where the two people are sufficiently distant from the drill they may speak normally since the apparent noise power from the drill is relatively modest. If they move towards the drill its apparent noise increases and, in order to compensate, they tend to speak louder. Eventually, when they reach the drill, it may be so loud that they may no longer hear each other over its noise. Clearly, where signal exceeds noise by an adequate margin communication may occur. Where signal and noise levels are similar, and certainly if noise power exceeds that of the signal, communication fails. Hence we see that it is the relative values of signal and noise powers, rather than their absolute values that determines whether overall performance is acceptable. The margin that would be suitable is dependent upon the type of system and is generally determined by means of subjective tests. Typical values of acceptable signal to noise ratio are 50–60 dB for high-quality music broadcast, 16 dB for low-grade speech and up to 30 dB for commercial telephony systems and about 60 dB for good quality TV transmission.

EXAMPLE 3.4
An amplifier has an output resistance of 600 Ω and delivers 250 mV into a matched load. If the temperature of the load resistor is 27°C and signal bandwidth 10 MHz, determine:

(a) available noise power;
(b) signal power developed in the load;
(c) signal to noise ratio at the load in dB.

SOLUTION

(a) Available noise power $= kT_sB$ $\qquad$ (3.48)

$$= 1.38 \times 10^{-23} \times 300 \times 10 \times 10^6 \qquad (3.49)$$

$$= 4.14 \times 10^{-14}\,W \qquad (3.50)$$

(b) Signal power in load $= \dfrac{V^2}{R_1}$ $\qquad$ (3.51)

$$= \frac{(250\,mV)^2}{600\,\Omega} \qquad (3.52)$$

$$= 0.1042\,mW \qquad (3.53)$$

(c) From Eqn (3.47) we may write:

$$\frac{S}{N}\,dB = 10\log\left(\frac{0.1042\,mW}{4.14 \times 10^{-14}\,W}\right) \qquad (3.54)$$

$$= 94\,dB \qquad (3.55)$$

Another advantage of expressing S/N in decibels rather than a ratio is that where two, or more, networks are cascaded, the overall S/N is the sum of the individual S/Ns, rather than the product which is more easily calculated. This is illustrated in Example 3.5.

EXAMPLE 3.5
Determine the S/N at the output of the system shown in Figure 3.10 as a ratio and in dB if the input signal is 2 mW accompanied by a noise level of 5 μW. Assume each network to be noiseless.

SOLUTION

$$S/N \text{ at the input, } (S/N)_{in} = \frac{2\,mW}{5\,\mu W} = 400 \qquad (3.56)$$

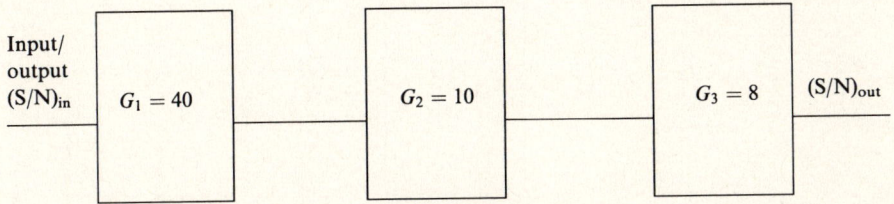

Figure 3.10

S/N at the output, $(S/N)_{out} = G_3 \times G_2 \times G_1 \times (S/N)_{in}$ (3.57)

$$= 8 \times 10 \times 40 \times 400 \qquad (3.58)$$

$$= 1.28 \times 10^6 \qquad (3.59)$$

The $(S/N)_{out}$ dB equals $(S/N)_{in}$ plus the gain of each stage in the system, where all quantities are in dB.

S/N at the input, $(S/N)_{in} = 10 \log_{10} \dfrac{2\,mW}{5\,\mu W}$ (3.60)

$$= 10 \log_{10} 400 \qquad (3.61)$$

$$= 26.02\,dB \qquad (3.62)$$

We may express the gain of each section in dB:

$$G_1 = 10 \log_{10} 40 = 16.02\,dB \qquad (3.63)$$

$$G_2 = 10 \log_{10} 10 = 10\,dB \qquad (3.64)$$

$$G_3 = 10 \log_{10} 8 = 9.03\,dB \qquad (3.65)$$

Hence the overall system gain is:

$$System\ gain = G_1 + G_2 + G_3 \qquad (3.66)$$

$$= 16.02\,dB + 10\,dB + 9.03\,dB = 35.05\,dB \qquad (3.67)$$

and finally $(S/N)_{out}$ dB $= (S/N)_{in}$ dB + system gain

$$= 26.02\,dB + 35.05\,dB \qquad (3.68)$$

$$= 61.07\,dB \qquad (3.69)$$

Self-assessment 3.3 In the system shown in Figure 3.11 determine:

(a)   the gain of the first stage;
(b)   the noise power at the input, and output of the second and third stages;
(c)   the signal power at the output of the first and third stages.

Assume that each stage is 'noiseless'.

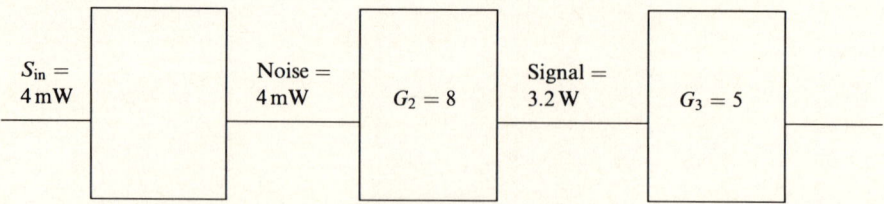

Figure 3.11

3.6 Noise figure

The input to a system, or sub-system, has a signal and associated noise and therefore has a certain input S/N, or $(S/N)_{in}$. As the signal, and noise, pass through a system they are both subjected to the same gain, or attenuation, of each stage. It might reasonably be expected that the output S/N, or $(S/N)_{out}$, be identical to that of the input which means that $(S/N)_{out}$ is neither worse or better than the value of $(S/N)_{in}$. In practice noise is introduced by every stage of a system such that noise progressively increases relative to the signal and hence S/N gradually deteriorates as a signal passes through a system.

The noise introduced by a system, or sub-system, may conveniently be indicated by its **noise figure**, F, which, providing the system is matched throughout, may be defined as:

$$F = \frac{\text{Input signal/noise ratio}}{\text{Output signal/noise ratio}} \qquad (3.70)$$

$$= \frac{(S/N)_{in}}{(S/N)_{out}} \qquad (3.71)$$

where $(S/N)_{out}$ and $(S/N)_{in}$ may either be ratios, or expressed in dB. Therefore the noise figure may be expressed as either a ratio of two ratios equating to a numeric value which is dimensionless or, where input and output S/N values are expressed in dB, has a value in unit dB.

Where S/N at a system's input is identical to that of the output, the system clearly introduces no noise and the noise figure is 1, or 0, dB. Where a system introduces additional noise, S/N at the input exceeds that of the output and the noise figure has a value greater than 1. Thus we may see how the noise figure is an indicator of the 'noisiness', or otherwise, of a system, sub-system or circuit.

The effective noise temperature T_e and noise figure, in ratio form, are related by the following expression:

$$T_e = T_0(F - 1) \qquad (3.72)$$

where T_0 is 290 K, the temperature commonly used for reference purposes in equipment specification.

EXAMPLE 3.6

A system has an input noise of 20 pW, output signal of 0.4 mW and output noise of 4 nW. If the gain of the system is 40, determine:

(a) the noise contribution of the system;
(b) the signal to noise ratio at the input;
(c) noise figure.

SOLUTION

(a) With a gain of 40, the noise contribution at the receiver due to input noise is:

$$= G \times N_{in} \tag{3.73}$$

$$= 40 \times 20 \, pW \tag{3.74}$$

$$= 800 \, pW \tag{3.75}$$

Therefore the noise contribution of the system is given by:

$$N_{out} - N_{in} \tag{3.76}$$

$$= 4 \, nW - 800 \, pW \tag{3.77}$$

$$= 3.2 \, nW \tag{3.78}$$

(b) We know the noise power at the input. In order to determine the input signal to noise ratio we must first find the signal power at the input:

$$\text{Input signal, } S_{in} = \frac{S_{out}}{G} \tag{3.79}$$

$$= \frac{0.4 \, mW}{40} \tag{3.80}$$

$$= 10 \, \mu W \tag{3.81}$$

$$\therefore \quad (S/N)_{in} = \frac{S_{in}}{N_{in}} \tag{3.82}$$

$$= \frac{10 \, \mu W}{20 \, pW} \tag{3.83}$$

$$= 500\,000 \text{ or } \simeq 57 \, dB \tag{3.84}$$

(c) Noise figure:

$$F = \frac{(S/N)_{in}}{(S/N)_{out}} \tag{3.85}$$

$$(S/N)_{in} = \frac{0.4 \, mW / 40}{20 \, pW} \tag{3.86}$$

$$= 500\,000 \tag{3.87}$$

$$(S/N)_{out} = \frac{0.4 \, mW}{4 \, nW} \tag{3.88}$$

$$= 100\,000 \tag{3.89}$$

Substituting Eqns (3.87) and (3.89) into (3.85):

$$F = \frac{500\,000}{100\,000} \qquad (3.90)$$

$$= 5 \text{ or } \approx 7\,\text{dB} \qquad (3.91)$$

As a final check we could check the calculation for $(S/N)_{in}$ in (c) above. The difference between $(S/N)_{in}$ and $(S/N)_{out}$, both in dB, should equal the noise figure calculated in part (b), again in dB. It is left to the reader to verify that it is in fact the case.

Self-assessment 3.4 The output of an antenna is connected to a pre-amplifier which has a noise figure of 8 and a gain of 17 dB. If the noise power in a given bandwidth at the antenna output is 12 pW calculate:

(a) the noise power in the given bandwidth at the amplifier output;
(b) the noise contribution of the amplifier.

Now consider a number of networks in cascade, each with either gain G, or loss L (note $L = 1/G$) and associated noise figures as shown in Figure 3.12. Determination of the overall noise figure F for the above system using Eqn (3.70) requires knowledge of signal to noise ratios. Alternatively F may be found by what is known as de Friis's equation, which is shown below for the general case of n cascaded sections:

$$F = F_1 + \frac{F_2 - 1}{G_1} + \frac{F_3 - 1}{G_1 G_2} + \cdots + \frac{F_n - 1}{G_1 G_2 \cdots G_n} \qquad (3.92)$$

where

$$G_2 = 1/L_2 \qquad (3.93)$$

G_1 and F_1 represent the gain and noise figure, respectively, of the first stage and G_2 and F_2 the second, and so on.

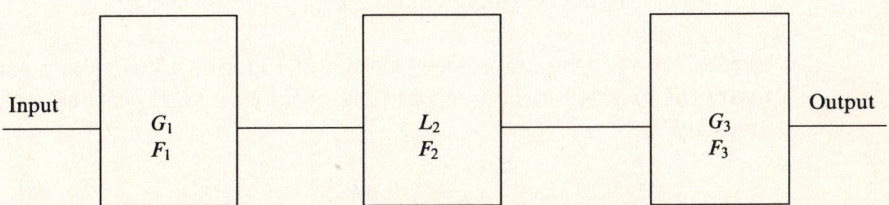

Figure 3.12 Noise figure of cascaded networks

An interesting and important point to note about de Friis's equation is that the overall noise figure F is critically dependent upon the first stage. Certainly, where the first stage has gain, its noise figure, and therefore contribution, is descaled by an amount equal to the first gain G_1. Subsequent stages, assuming all stages have gain rather than loss, have less effect upon the overall noise figure because subsequent denominator terms are gain *products*. Such products generally are of such a high value that, when divided, the noise contribution may then be ignored in relation to that of F_1.

A common decision faced by engineers is where to place an amplifier in a radio receiving system. Radio signals that are received are invariably very weak and amplification is required to produce an adequate voltage from an associated antenna and feeder. From a casual consideration it may not at first seem to matter whether an amplifier is placed between antenna and feeder or at the feeder output. Certainly it makes no difference in relation to signal power received. Consider the following example.

EXAMPLE 3.7

An antenna is connected to a receiver by a feeder cable which has a loss of 3 dB and noise figure 0.25 dB. A pre-amplifier, with gain of 18 dB and noise figure 1 dB, is to be connected prior to the receiver. Assuming matched conditions, compare the effect upon noise figure of positioning the pre-amplifier:

(a) between antenna and feeder;
(b) between feeder and receiver.

SOLUTION

Before determining the noise figure F of the antenna and feeder networks in cascade by means of de Friis's equation, it is necessary to express the gain and noise figures as a ratio, rather than in the dB form given.

$$0.25\,\text{dB is equivalent to } a\log_{10} 0.025 = 1.059$$

$$1\,\text{dB is equivalent to } a\log_{10} 0.1 = 1.259$$

$$3\,\text{dB is equivalent to } a\log_{10} 0.3 = 1.995$$

$$18\,\text{dB is equivalent to } a\log_{10} 1.8 = 63.096$$

(a) Feeder followed by amplifier is shown in Figure 3.13(a). The noise figure for two networks in cascade is, from Eqn (3.92) earlier, given by:

$$F = F_1 + \frac{F_2 - 1}{G_2} \tag{3.94}$$

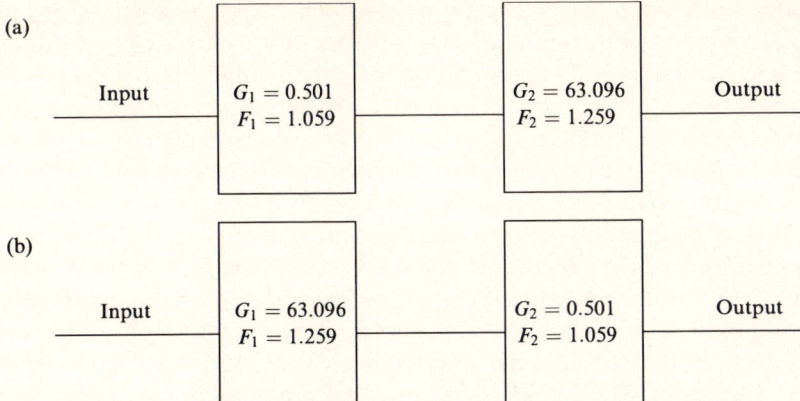

Figure 3.13   (a) Feeder followed by amplifier. (b) Amplifier followed by feeder

In the above configuration, substituting values for gain and noise figure:

$$F = 1.059 + \frac{1.259 - 1}{63.096} \tag{3.95}$$

$$= 0.266\,\text{dB} \tag{3.96}$$

(b) Amplifier followed by feeder is shown in Figure 3.13(b). Now with the amplifier and feeder reversed, and again substituting values of gain and the noise figure:

$$F = 1.259 + \frac{1.059 - 1}{0.501} \tag{3.97}$$

$$= 1.389\,\text{dB} \tag{3.98}$$

Thus it may be seen that if the amplifier precedes the feeder, the noise figure is reduced in this example from 1.389 dB to 0.266 dB, an appreciable improvement which improves the output S/N ratio by 1.123 dB.

3.7 Summary

Electrical noise may be defined as any unwanted energy that accompanies a signal in a communication system. A signal at any point within a system is always accompanied by some noise. Noise in communication systems may be categorised as artificial or as occurring naturally. Many electrical and electronic components naturally introduce noise into a system.

A major source of noise is thermal noise which occurs in resistors because of thermal agitation and is proportional to temperature and the bandwidth of operation. Thermal noise, and a number of other noise sources that occur in electronic components, may be modelled, enabling their noise power contributions to be calculated in electrical circuits. Another model to represent noise introduced by an electrical network makes use of the concept of effective noise temperature. The noise power in a load purely due to the contribution of a network may be found by knowledge of its source resistance and the effective noise temperature of the network, in conjunction with the gain of the network. Noise peculiar to radio systems may occur naturally due to static, quantum or earth noise as well as cosmic and solar noise. A particular problem in some radio systems is that of artificial interference produced by certain types of electrical equipment. The combined effects of naturally occurring noise appear at the input to a receiving antenna and may be represented as an equivalent noise temperature of the antenna and typical values are presented. The concept of effective noise temperature may be used to represent the noisiness of a receiver by expressing it in terms of a noise temperature at its input. This is a convenient point to express noise temperature because different receivers may then be directly compared.

The concept of signal to noise ratio was introduced and it was demonstrated that even in the presence of noise a system may perform satisfactorily providing there is a reasonable margin of signal power over that of noise. The noise figure of a network provides an alternative approach to the use of the signal to noise ratio. The noise figure enables overall noise performance and the signal to noise ratio to be readily calculated for a system comprising a number of networks in cascade.

Exercises

3.1   Determine an equivalent noise circuit to represent two resistors R_1 and R_2 connected in series at temperatures T_1 and T_2 respectively. Hint: Use Thévenin's theorem to determine two independent Thévenin electromotive forces (emfs), e_1 and e_2. Because these emfs are rms, the emf of the equivalent noise circuit e_n may be found thus:

$$e_n = \sqrt{(e_1^2 + e_2^2)} \tag{3.99}$$

3.2   A diode is connected in series with a resistance of $4.7\,\text{k}\Omega$. If the mean value of current is $2\,\text{mA}$, determine the noise voltage developed in the resistor as a result of both shot and thermal noise. Assume a bandwidth of $100\,\text{kHz}$ and an ambient temperature of $27°\text{C}$.

3.3 All stars within the galaxy emit RF radiation. In considering cosmic noise suggest a reason why only solar noise is considered significant. (Note: attenuation of an electromagnetic wave is proportional to the square of distance.)

3.4 A system comprises a stage with a gain of 100 and effective noise temperature of 600°C followed by another stage with gain of 500 and effective noise temperature of 3000°C. The signal input to the system is accompanied by 12 pW of noise power in a bandwidth of 2 MHz. Assuming that the system is matched throughout, calculate:

    (a) the effective noise temperature of both stages, referenced to the system input;
    (b) the noise contribution of each stage;
    (c) the total noise power at the output of the system.

3.5 With reference to Figure 3.14 determine the signal power and noise power at each point in the system and the gain of the second stage. Assume that each stage does not contribute any noise and that the system is matched throughout.

3.6 The output of an antenna is connected to a pre-amplifier which has a gain of 15 dB. If the noise power in a given bandwidth at the input to the amplifier is 12 pW and the noise power at the output is 3 nW calculate:

    (a) the noise power at the amplifier output due to the input noise;
    (b) the noise contribution of the amplifier;
    (c) the noise figure of the amplifier;
    (d) the effective noise temperature at the input to the amplifier.

3.7 A system comprises three amplifiers in cascade, each of different gain. Explain why, in qualitative terms, the signal to noise ratio at the system output is a maximum if the stages are arranged such that the gain of each successive stage is less than that of its predecessor.

3.8 A 50 Ω satellite receiver has a bandwidth of 10 MHz. The receiver consists of an antenna with effective noise temperature of 20 K followed by a pre-amplifier with gain of 30 dB and an effective noise temperature of 4 K.

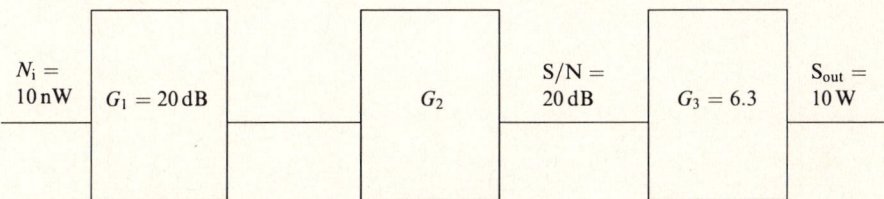

Figure 3.14

The feeder, which has a loss of 1.5 dB, is connected to another stage having a gain of 65 dB and noise figure of 15 dB. If the ambient temperature is 17°C, determine:

(a) the noise figure of the pre-amplifier;
(b) the effective noise temperature of the feeder;
(c) the noise figure of the system, ignoring the antenna;
(d) the effective noise temperature of the entire system at the input to the pre-amplifier.

Bibliography

Connor, F.R., *Noise*, 2nd edn, Edward Arnold, 1982. ISBN 0-7131-3459-3.
Schwartz, M., *Information Transmission, Modulation and Noise*, 4th edn, McGraw-Hill, 1990. ISBN 0-07-100931-0.

CHAPTER 4

Modulation

Aims and objectives

The distinction between amplitude modulation and angle modulation is made. Angle modulation may be categorised as either phase or frequency modulation. Phase modulation is far less common and the chapter concentrates upon amplitude and frequency modulation.

A detailed analysis is presented of full amplitude modulation and its time and frequency representations. Modulation depth is defined and its influence upon noise immunity and power distribution in carrier and side frequency components. Non-linear and linear methods of amplitude modulation production are described as well as the use of a diode detector, or a coherent technique, for demodulation. Suppressed carrier and single sideband variants of amplitude modulation are introduced. A mathematical representation of frequency modulation is given and the modulation index defined. The spectrum of a frequency modulated signal is developed to show that it is theoretically infinite. The use of Bessel functions is demonstrated to determine a suitable practical, finite, bandwidth based upon modulation index. Alternatively, Carson's rule may be used to approximate the bandwidth required.

Narrowband frequency modulation is introduced and contrasted with both amplitude and frequency modulation. An example of frequency division multiplexing is offered to illustrate a common technique of multi-channel operation over a single transmission path. Threshold effect is explained.

Demodulation of signals in the presence of noise is examined and the signal to noise ratio at the output of a receiver estimated. A detailed comparison, principally in relation to noise performance and bandwidth requirement, is made for the various modulation techniques, as well as comparison with equivalent baseband operation.

4.1 Amplitude modulation

Modulation may be defined as the process of impressing information, in the form of a **message signal $m(t)$**, upon another waveform of higher frequency known as a **carrier**. One of the principal reasons for employing modulation is to convert signals of lower frequency into sufficiently high frequency for transmission at radio frequencies. This may be to make use of radio communication or to facilitate multi-channel operation in cable systems, be they metallic or optical fibre, where each channel operates within a unique spectral band in order to avoid interference with each other.

A carrier, here assumed sinusoidal, may be represented by the general expression:

$$g(t) = A \sin(\omega t + \phi) \tag{4.1}$$

where: ω is the angular velocity and
         ϕ is an arbitrary value of phase.

The amplitude A could be modulated in some way to produce **amplitude modulation**, or AM. Equation (4.1) could alternatively have been expressed:

$$g(t) = A \sin \theta(t) \tag{4.2}$$

where $\theta(t)$ is an angle that is in turn a function of time and, in comparing with Eqn (4.1):

$$\theta(t) = \omega t + \phi \tag{4.3}$$

The angle $\theta(t)$ may also be modulated as an alternative to the amplitude of the carrier. **Angle modulation** may be produced in one of two ways, viz. **phase modulation** (PM) where ϕ is modulated such that its instantaneous phase is proportional to the amplitude of the message signal, or **modulating signal**, or **frequency modulation** (FM) where frequency is changed in proportion to the instantaneous amplitude of the modulating signal.

Amplitude modulation may be simply expressed mathematically as:

$$g(t) = \left(1 + \frac{V_m}{V_c} \sin \omega_m t\right) V_c \sin \omega_c t \tag{4.4}$$

where $V_c \sin \omega_c t$ represents a carrier waveform onto which a modulating signal $V_m \sin \omega_m t$ may be impressed. By means of trigonometrical identities for the expansion of the product of two sine terms, $g(t)$ may be rewritten:

$$g(t) = V_c \sin \omega_c t + \frac{V_m}{2} \cos(\omega_c - \omega_m)t - \frac{V_m}{2} \cos(\omega_c + \omega_m)t \tag{4.5}$$

Equation (4.5) contains a sinusoidal component $V_c \sin \omega_c t$ which is simply the unmodulated carrier. To this is added what is known as the sum and difference components, $V_m \sin(\omega_c + \omega_m)t$ and $V_m \sin(\omega_c - \omega_m)t$, called **upper** and **lower side frequencies**, respectively. These appear, in double-sided form, as two

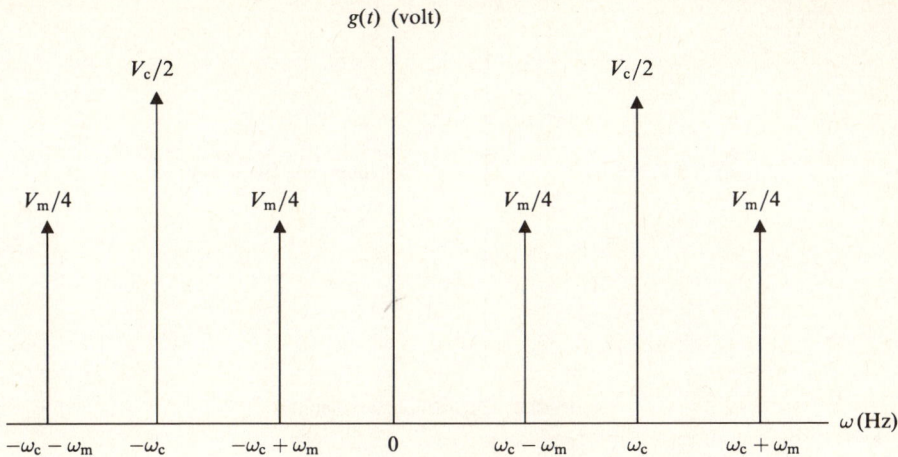

Figure 4.1 Full AM spectrum

sinusoids spaced at $\pm(\omega_c \pm \omega_m)$. The signal is known as a **full AM** signal and its spectrum is shown in Figure 4.1 which is equivalent to the convolution of ω_c and ω_m.

In practice a modulating signal is a complex wave and therefore consists of a range of frequencies known as a **band** which represents the original information. It is common to show an AM signal with such a modulating or **baseband** signal as shown below in Figure 4.2, where f_u is the highest, or upper cut-off, frequency of the message signal.

It may be seen that an AM signal has a bandwidth twice that of the highest frequency component of the baseband signal. The information that is to modulate a carrier must be of much lower frequency than that of the carrier in order that aliasing, discussed in Chapter 2, does not occur. From Figure 4.1 it is evident that $\omega_c \geqslant 2\omega_m$ to prevent overlap of positive and negative frequencies.

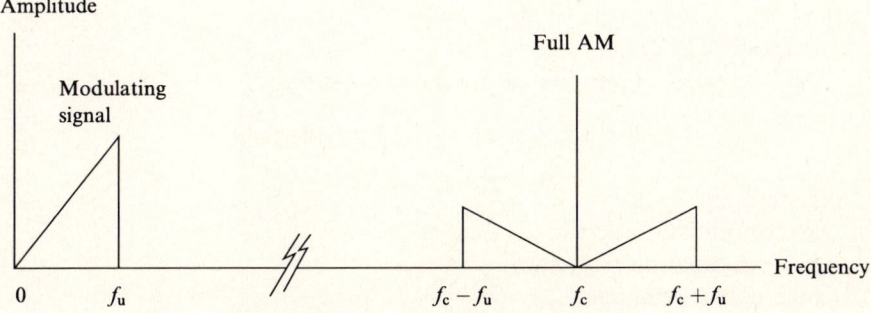

Figure 4.2 Full AM spectrum (in practice)

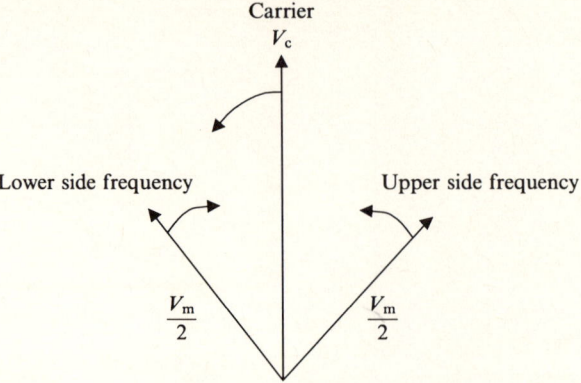

Figure 4.3 Full AM phasor diagram

Self-assessment 4.1 A telephone speech signal has a bandwidth ranging from 300 to 3400 Hz and is used to modulate a carrier frequency of 120 kHz. Sketch the double-sided spectrum of the modulated signal.

Before examining AM in the time domain it is helpful to consider the phasor diagram representation of $g(t)$, Figure 4.3. From the figure it may be realised that the time domain representation is a sinusoid of angular velocity equal to that of the carrier frequency whose peak amplitude ranges between $V_c \pm V_m$, hence the name amplitude modulation. A sketch of $g(t)$ is shown in Figure 4.4. The ratio V_m/V_c is known as **modulation depth** m:

$$m = \frac{V_m}{V_c} \tag{4.6}$$

Equation (4.6) may be substituted into Eqn (4.4) to yield an alternative, and more common, expression for an AM waveform:

$$g(t) = (1 + m \sin \omega_m t) V_c \sin \omega_c t \tag{4.7}$$

EXAMPLE 4.1
An AM wave is represented by the expression:

$$e = 5(1 + 0.5 \cos 3140t) \sin 2\pi 10^5 t$$

Determine:
(a) modulation depth;
(b) modulating frequency;
(c) carrier frequency;
(d) peak instantaneous amplitude of the modulated wave.

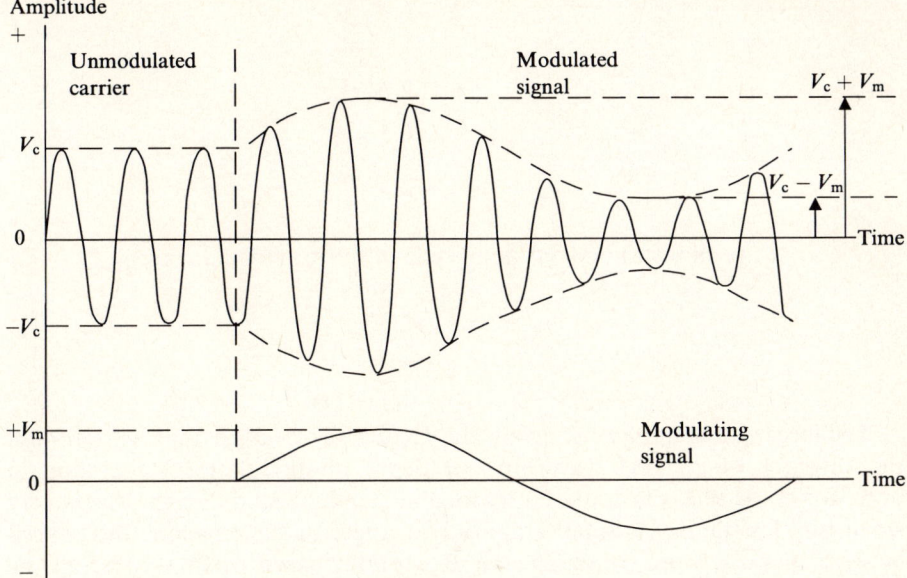

Figure 4.4 Full AM (time domain)

SOLUTION
The expression above indicates:

$$V_c = 5\,\text{V}$$

and that carrier is expressed by $5\sin^2 \pi 10^5\,\text{V}$. Also

$$\frac{V_m}{V_c}\sin \omega_m t = 0.5\cos 3140t\,\text{V}$$

(a) by inspection $m = 0.5$ and

(b) $\omega_m = 3140$

$$\therefore \quad f_m = \frac{3140}{2\pi} = 500\,\text{Hz}$$

(c) $\omega_c = 2\pi 10^5$

$$\therefore \quad f_c = \frac{2\pi 10^5}{2\pi} = 100\,\text{kHz}$$

(d)
$$\frac{V_m}{V_c} = 0.5$$

$$\therefore \quad V_m = 0.5 V_c$$

$$= 0.5 \times 5\,\text{V}$$

$$= 2.5\,\text{V}$$

$\therefore$ peak instantaneous amplitude of the modulated wave $= V_c + V_m$

$$= 5 + 2.5\,\text{V}$$

$$= 7.5\,\text{V}$$

The larger the modulating signal, the greater the peak to peak variation in the envelope of an AM waveform, or depth of modulation. Operation in practice should seek to produce a reasonable depth of modulation. If m is very small, say less than 0.1, small amplitude changes in the envelope introduced because of noise may be significant. Reception, based upon recovering the envelope, produces the required message signal and the noise accompanying the AM signal, leading to a reduced S/N ratio. Alternatively if m is sufficiently large, the negative peak of the envelope may touch the horizonal axis. Should the modulating signal be further increased, the modulated waveform distorts and the demodulated signal will no longer be true. In addition, for large m, when the minimum peak to peak amplitude of the modulated signal is small, noise again becomes significant.

For a practical AM modulator, there is thus a limit to how small, or large, a modulating signal may be for a given carrier amplitude if unacceptable noise and distortion is to be avoided. Fulfilling the above requirements is particularly difficult where a signal has a large dynamic range. Typically modulation depth should remain within 0.2 to 0.8 for 99% of the time to ensure satisfactory operation.

Equation (4.6) may be conveniently modified into the form shown in Eqn (4.8) which is particularly useful for determining m in practice from an AM waveform displayed upon an oscilloscope:

$$m = \frac{(\text{maximum} - \text{minimum})\ \text{amplitude of envelope}}{(\text{maximum} + \text{minimum})\ \text{amplitude of envelope}} \tag{4.8}$$

$$= \frac{(V_c + V_m) - (V_c - V_m)}{(V_c + V_m) + (V_c - V_m)} \tag{4.9}$$

Self-assessment 4.2 The envelope of an AM waveform has a maximum and minimum peak amplitudes of 14 V and 6 V, respectively. Determine:

(a) modulation depth;
(b) carrier amplitude;
(c) amplitude of the modulating signal.

Consider the power developed by a modulated signal into a load resistor R:

$$\text{Power in load} = \frac{V^2}{R} \tag{4.10}$$

$$\therefore \quad \text{Carrier power, } P_c = \frac{V_c^2}{R} \tag{4.11}$$

From Eqn (4.6) we may write:

$$V_m = mV_c \tag{4.12}$$

Hence the power developed in the load due to each sideband is given by:

$$\frac{(mV_c/2)^2}{R} \tag{4.13}$$

$$= \frac{m^2 V_c^2}{4R} \tag{4.14}$$

$$= \frac{m^2}{4} P_c \tag{4.15}$$

Therefore the total modulated power developed in R is:

$$P = \frac{V_c^2}{R}\left(1 + 2\frac{m^2}{4}\right) \tag{4.16}$$

$$= \frac{V_c^2}{R}\left(1 + \frac{m^2}{2}\right) \tag{4.17}$$

The ratio of power in the sidebands to total power equals:

$$\frac{2m^2 V_c^2/4R}{\dfrac{V_c^2}{R}\left(1 + \dfrac{m^2}{2}\right)} \tag{4.18}$$

$$= \frac{m^2}{2 + m^2} \tag{4.19}$$

Alternatively the following expression for the ratio of carrier power to total power may be developed:

$$\frac{P_c}{P} = \frac{2}{2 + m^2} \tag{4.20}$$

We may see from Eqn (4.20) that a significant proportion of the total power developed in R is purely due to the carrier and makes up upwards of two-thirds of the total power. This is wasteful of transmitter power, especially in high-power applications such as television broadcasting, and is a potential disadvantage of full AM.

Self-assessment 4.3 Determine the percentages of sideband and carrier power developed in a load resistance for values of m between 0 and 1, in 0.1 increments. Show your results in the form of a graph.

Amplitude modulation is performed by either an electronic multiplier or a mixer. Although different in operation both achieve the effect of multiplying V_c and V_m together which is a necessary condition as indicated in Eqn (4.4) earlier.

4.1.1 Non-linear modulation

Non-linear modulation arranges to sum carrier and modulating signal. The resultant waveform is then swept sufficiently widely across the transfer characteristic of a device, e.g. an amplifier, to ensure operation encompasses the non-linear region. This process is known as **mixing**. The transfer function of a device operating in a non-linear mode relating output to input is of the general form:

$$V_o = k + aV_i + bV_i^2 + cV_i^3 + \cdots \tag{4.21}$$

where V_o and V_i represent output and input voltages.

Now, suppose that V_i is the sum of carrier and modulating signal:

$$V_i = V_m \sin \omega_m t + V_c \sin \omega_c t \tag{4.22}$$

Then substituting Eqn (4.22) into (4.21) yields:

$$V_o = k + a(V_m \sin \omega_m t + V_c \sin \omega_c t) + b(V_m \sin \omega_m t + V_c \sin \omega_c t)^2 \tag{4.23}$$

Expansion of Eqn (4.23) includes the components:

$$aV_c \sin \omega_c t + bV_m V_c [\cos(\omega_c - \omega_m)t - \cos(\omega_c + \omega_m)t] \tag{4.24}$$

which will be recognised as a form of mathematical representation of an AM wave. All other components of the expansion of Eqn (4.23) (baseband frequency and multiples of the carrier frequency) may be rejected by suitable filtering. It is the use of the square term of the transfer characteristics of a device used for generation of AM that such modulators are also known as **square law** devices.

A disadvantage of non-linear modulation is that operation in practice is less than perfect, for example because of inadequate filtering, which results in a modulated waveform which is not necessarily linear and hence distorted. Typically, positive and negative half-cycles of the modulated waveform are not symmetrical.

4.1.2 Linear modulation

Where a modulated signal is required with a large power, e.g. high-power broadcast transmitters, and good linearity, **linear** modulation is necessary to reduce distortion. The modulation process makes use of a linear characteristic of a transfer function and a typical circuit found in practice is shown in Figure 4.5.

A common-emitter transistor amplifier is biased in Class C, that is in cut-off (base bias not shown). A parallel LC load, which is tuned to the carrier frequency and therefore purely resistive, provides an ac load line, of constant slope, as shown in Figure 4.6. The RF carrier input voltage is of constant amplitude and sufficiently large to sweep the transistor between cut-off and saturation, where V_{ce} 'bottoms', once every cycle. This gives rise to pulses of collector current at carrier frequency. The effect of the tuned resonant LC load is to reject frequency components distant from the carrier frequency resulting in an

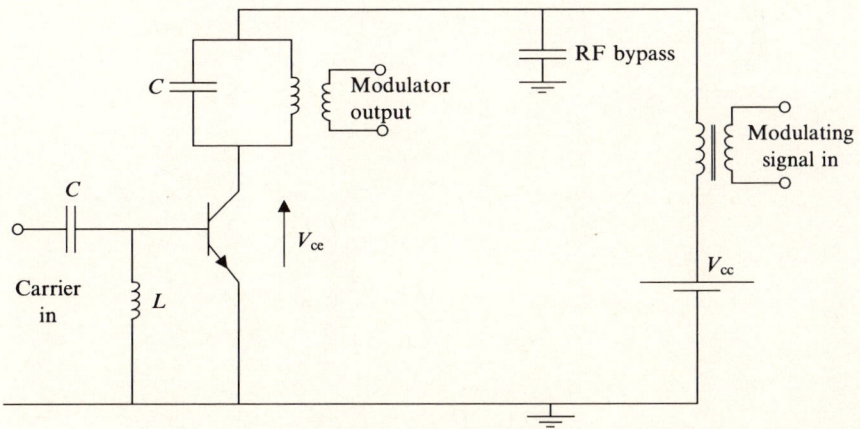

Figure 4.5   Linear modulator

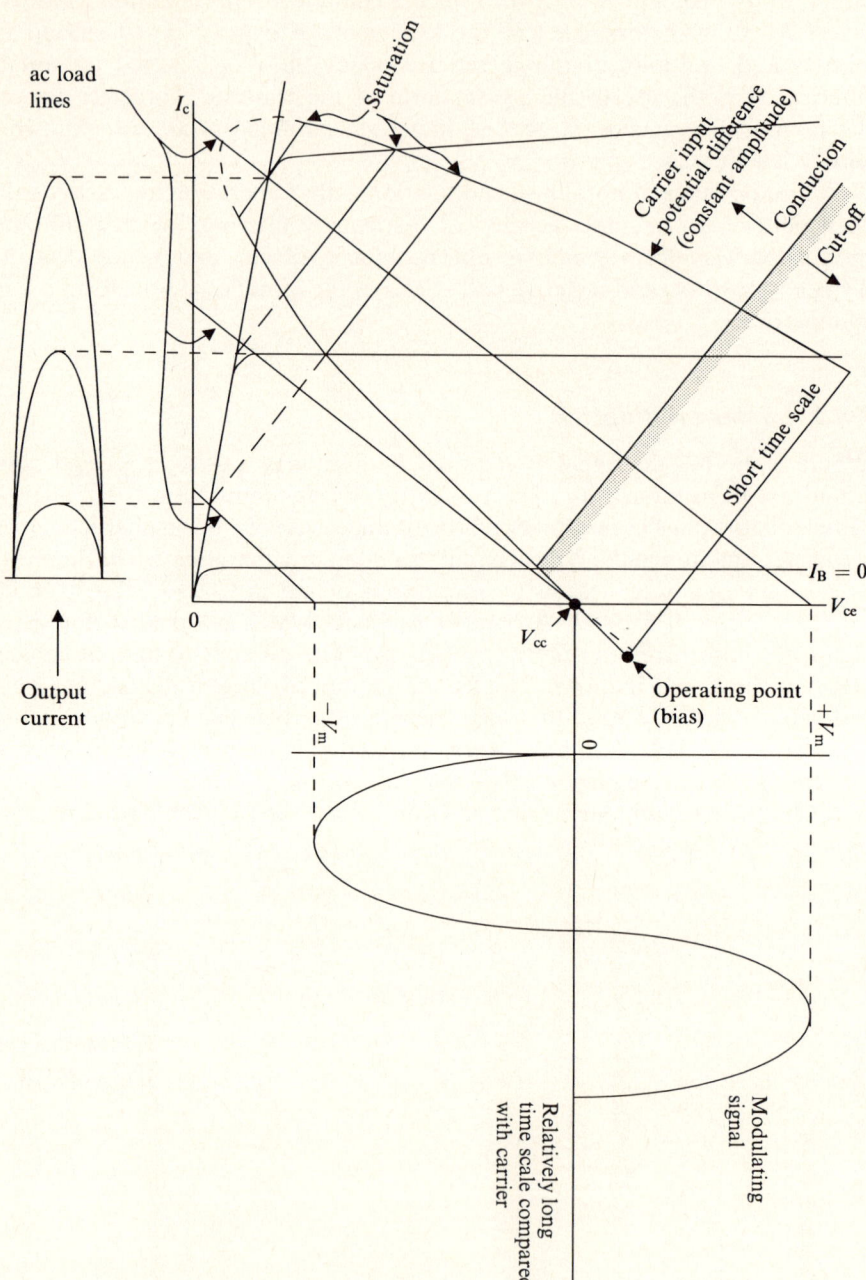

Figure 4.6   Transistor output characteristic

output voltage which, in the absence of modulation, is sinusoidal (remember, a common-emitter amplifier provides phase inversion of voltage).

The modulating signal V_m is added to the power supply voltage of the amplifier V_{cc} (hence the circuit is known as a modulated power amplifier) and gradually, in comparison to the rate of variation in carrier amplitude, sweeps the load line across the amplifier's output characteristic. This has the effect of modulating the peak value of I_c, and hence the output voltage V_{ce} which may be expressed:

$$V_{ce} = (V_{cc} + V_m) - V_{ce_{min}} \tag{4.25}$$

where $V_{ce_{min}}$ is the bottoming voltage which, as indicated above, is reached on every positive half-cycle of carrier for all values of V_m. Providing the bottoming line is *linear* and passes through the origin of the output characteristic, we may state that:

$$V_{ce_{min}} \propto (V_{cc} + V_m) \tag{4.26}$$

Hence we may write:

$$V_{ce_{min}} = k(V_{cc} + V_m) \tag{4.27}$$

where k is simply a constant of proportionality, the value of which depends upon the parameters of the circuit.

Now substituting Eqn (4.27) into (4.25):

$$V_{ce} = (V_{cc} + V_m) - k(V_{cc} + V_m) \tag{4.28}$$

$$= V_{cc} + V_m - kV_{cc} - kV_m \tag{4.29}$$

$$= (V_{cc} + V_m)(1 - k) \tag{4.30}$$

The output waveform of the circuit is an alternating current (ac) wave at carrier frequency whose envelope is governed by Eqn (4.30). A sketch of the output is shown in Figure 4.7 where it may be seen that a full AM waveform is produced similar to that shown earlier in Figure 4.4. The circuit produces a reasonably undistorted signal with amplifier efficiency of around 80%.

4.1.3 Demodulation of full AM

If a full AM signal is multiplied by its carrier frequency component part of the resulting spectral elements will contain the original message signal $m(t)$. One method of achieving the necessary multiplication is by application of the modulated signal to a square-law device. The square term of the transfer function will cross multiply a component at carrier frequency with that of the

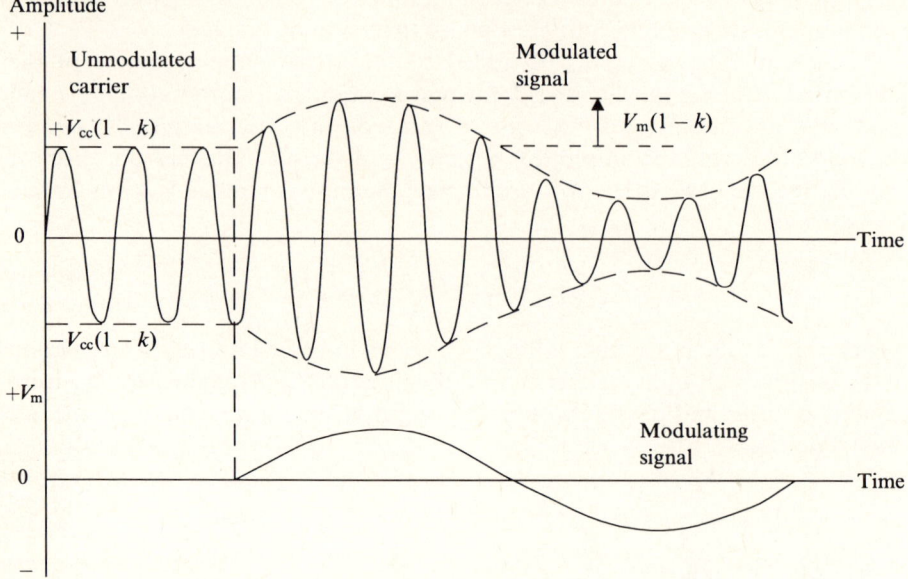

Figure 4.7   Modulated power amplifier output waveform

modulated signal. This will result in a baseband component. Note that unlike some later modulation forms, no separate carrier component is necessary in addition to that of the modulating signal.

The most common square law device used for demodulation of full AM is a signal diode. In order to use it for demodulation in the manner described above, the diode is operated as a small signal device, that is, it remains conducting all the time.

Figure 4.8 illustrates an alternative method for demodulating a full AM signal where a diode is used in large signal mode. The input voltage sweeps across the transfer characteristic to such an extent that the device continually cycles into, and out of, conduction. The circuit responds to the envelope (positive or negative half-cycles dependent upon diode connection) of the received signal and for this reason is known as an **envelope detector**.

Operation is described as though the diode is perfect in that its resistance is infinite when reverse biased and zero when forward biased. During positive half-cycles of the modulating signal the diode is on and the capacitor voltage V_c equals v_{in}. The time constant CR is made large in comparison to the period of the modulated signal. When the input voltage begins to fall, after having peaked, the capacitor voltage is unable to discharge through the diode and, because of the large time constant, is unable to follow the fall in input voltage. Rather, it discharges relatively slowly through resistor R. The voltage V_c is shown in Figure 4.9.

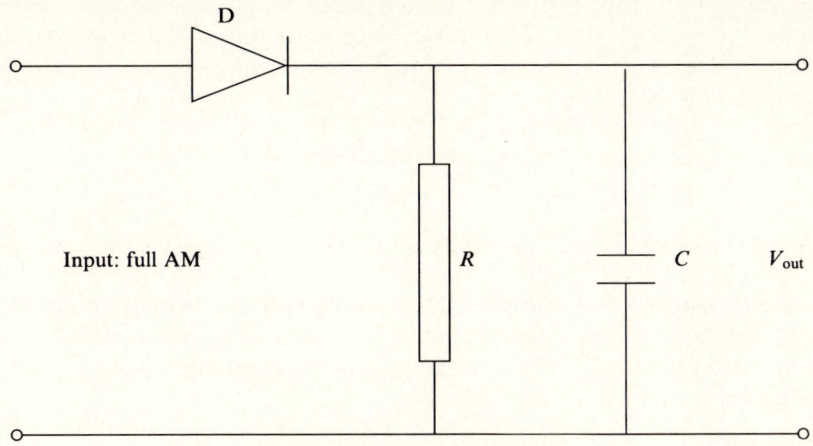

Figure 4.8 Diode detector

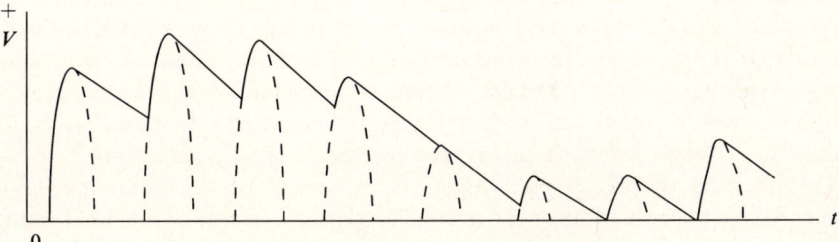

Figure 4.9 Output voltage of diode detector

With subsequent low-pass filtering to remove the high-frequency ripple, a demodulated signal is recovered which faithfully corresponds to that of the envelope which is, of course, equivalent to the modulating signal. If the time constant is made too large, a phenomenon known as **negative peak clipping** can occur which gives rise to distortion, Figure 4.10.

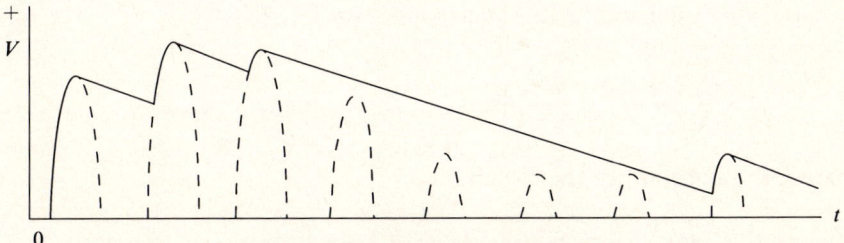

Figure 4.10 Negative peak clipping

The value of the time constant required depends upon the rate of change of envelope voltage with time. This in turn is governed by modulation depth m and also the frequency of the modulating signal. Connor[1] shows that the time constant should ideally be:

$$RC \leqslant \frac{\sqrt{(1 - m^2)}}{m\omega_m} \qquad (4.31)$$

The above relationship may generally be satisfied for smaller values of m.

Self-assessment 4.4 A carrier of 5 V is modulated by a signal of amplitude 2 V and frequency 4 kHz. If a diode detector is used for recovery of the modulating signal using a 10 kΩ resistor, estimate a suitable value of capacitance.

4.1.4 Balanced modulation

Full AM has the advantage that a simple, and cheap, diode detector circuit may be employed at the receiver. However, appreciable power is transmitted at the carrier frequency which is wasteful in that it contains no information energy. Another form of AM uses a **balanced** modulator which is characterised by the fact that no carrier is produced, simply lower and upper sidebands. This is known as **double sideband suppressed carrier** (DSBSC) operation.

One method of producing DSBSC is by means of a Cowan modulator, Figure 4.11. One half-cycle of the carrier produces a positive voltage at point A and a negative voltage at point B which forward biases all four diodes. This short-circuits the coils of the two transformers in the diode bridge circuit and hence no voltage occurs at the output. The other half-cycle of carrier reverse biases the diodes, open-circuiting the bridge, so allowing the modulating signal to pass to the output. The output waveform is equivalent to multiplying the modulating signal by positive half-cycles of an RF carrier, as shown in Figure 4.11(b).

Analysis of the Cowan modulator is performed by multiplying the modulating signal by the Fourier series representation of the carrier:

$$g(t) = kV_m \cos \omega_m t \left[\frac{1}{2} + \frac{2}{\pi} \left(\sin \omega_c t + \frac{1}{3} \sin 3\omega_c t + \cdots \right) \right] \qquad (4.32)$$

$$= \frac{kV_m \cos \omega_m t}{2} + \frac{kV_m}{\pi} [\sin(\omega_c - \omega_m)t + \sin(\omega_c + \omega_m)t] + \cdots \qquad (4.33)$$

where k is a constant of the circuit.

---

[1] Connor, F.R., *Modulation*, 2nd edn, Edward Arnold, 1982, pp. 80–81. ISBN 0-7131-3457-7.

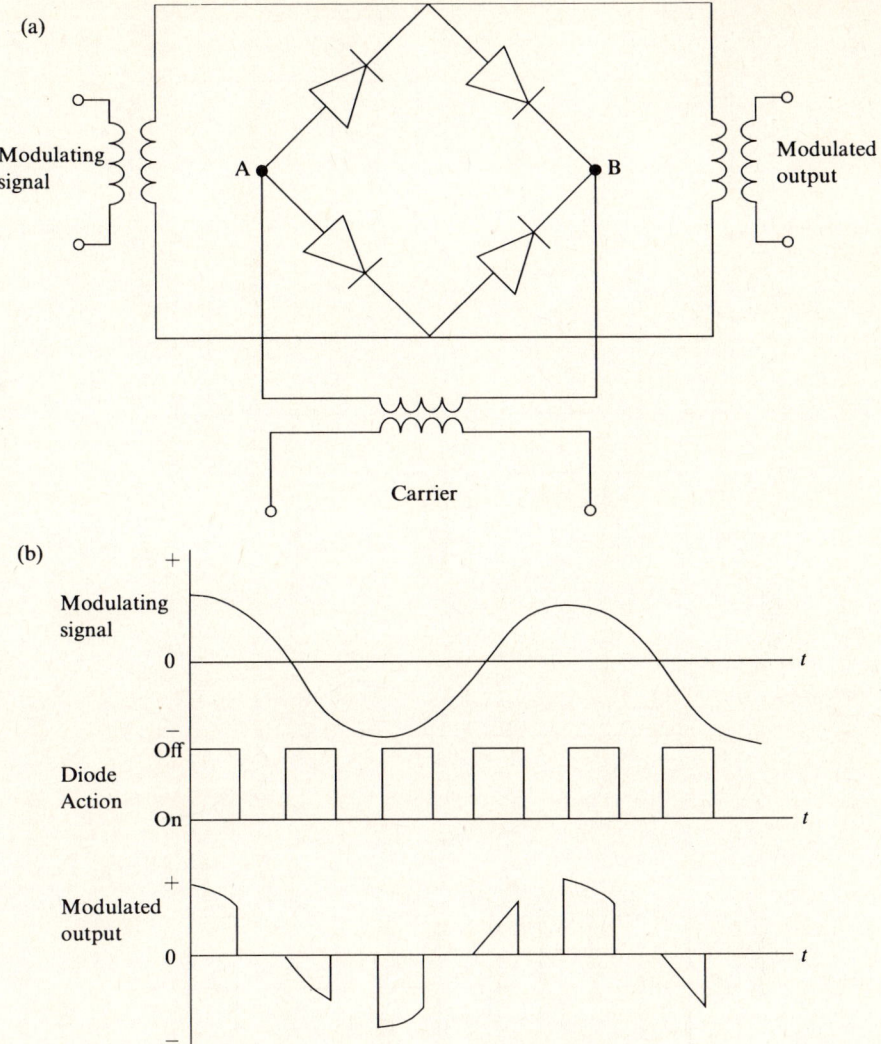

Figure 4.11 Cowan modulator: (a) circuit; (b) output waveform

Equation (4.33) contains a baseband component, upper and lower side frequency components which represents a DSBSC waveform and higher order frequency components centred upon odd multiples ω_c. There is no component at the carrier frequency. Band-pass filtering enables selection of the desired DSBSC frequency components and rejection of all others.

Figure 4.12 shows a Ring modulator where four diodes are arranged as a 'ring'. Dependent upon the half cycle of carrier frequency, either D1 and D2 are conducting, and D3 and D4 off, or vice versa. The effect is to pass the

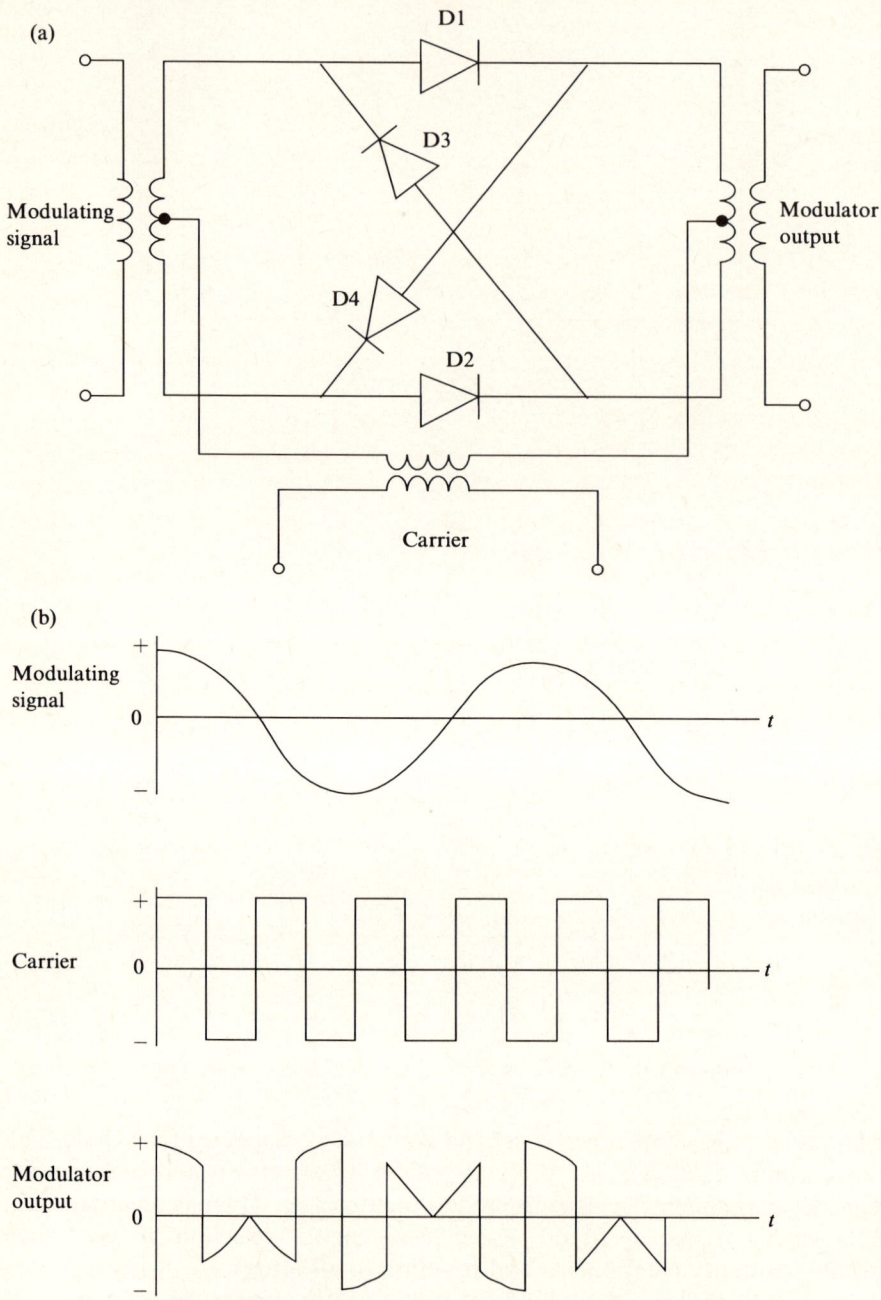

Figure 4.12   Ring modulator: (a) circuit; (b) output waveforms

instantaneous value of modulating signal directly to the output if D1 and D2 are on, or in inverted form if D3 and D4 are on, and results in the waveform shown in Figure 4.12(b). The diode ring arrangement simply acts as a reversing switch with switching frequency equal to that of the carrier frequency. Analysis is similar to that for the Cowan modulator but the square wave is bipolar, rather than unipolar. Again, a DSBSC expression is produced but it has twice the energy of the earlier chopped waveform found in the Cowan modulator.

Self-assessment 4.5 Analyse mathematically the ring modulator to produce an expression for the modulated output signal.

An envelope detector is unable to recover a DSBSC signal using a diode detector because, as seen in Figures 4.11(b) and 4.12(b), the envelope shape is unsuitable. Reception is performed by means of another balanced modulator. The received modulated signal is multiplied by a version of the original carrier which must be correctly phased with respect to the received DSBSC signal. This is known as **coherent detection**. In order to produce the carrier at the receiver two possible techniques are:

1. Transmission of the carrier over a separate path.
2. Reinsertion of the carrier at the transmitter, at reduced level, known as a **pilot carrier**. This offers considerable power saving compared with full AM and does not require a separate carrier path.

Reception deteriorates if the phase of the receiver's carrier drifts. Consider coherent detection of a DSBSC signal using a carrier of correct frequency, but arbitrary phase ϕ as shown in Figure 4.13.

$$v_{out} = AB \cos \omega_m t \cos \omega_c t \cos(\omega_c t + \phi) \tag{4.34}$$

$$= c/2 \cos \omega_m t [\cos(\omega_c t - \omega_c t - \phi) + \cos(\omega_c t + \omega_c t + \phi)] \tag{4.35}$$

$$= c/2 \cos \omega_m t [\cos \phi + \cos(2\omega_c t + \phi)] \tag{4.36}$$

$$= c/2 \cos \omega_m t \cos \phi + c/2 \cos \omega_m t \cos(2\omega_c t + \phi) \tag{4.37}$$

where c is simply a constant.

Equation (4.37) indicates that the first term of the expression is the desired demodulated signal, but is multiplied by $\cos \phi$. In addition there are a pair of side frequencies centred upon a frequency equal to twice that of the carrier. These may be removed by means of a low-pass filter. The demodulated signal amplitude diminishes as the receiver's carrier phase ϕ increases such that for $\phi = 90°$, it disappears completely, rendering detection impossible. In practice care is therefore necessary to ensure correct carrier phasing in coherent detectors.

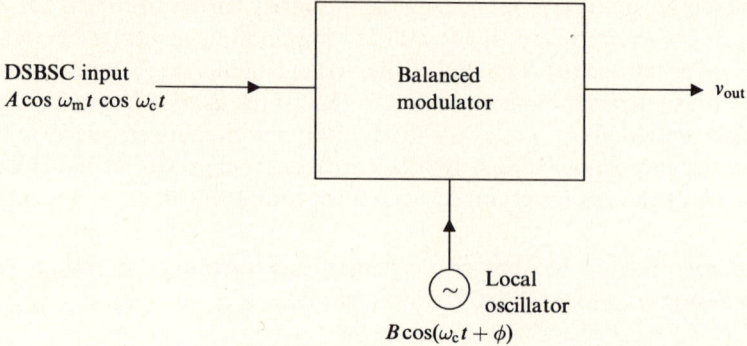

Figure 4.13   Coherent detection

Each sideband in full AM contains the full information content of the baseband signal. From an information consideration satisfactory transmission may be achieved using **single sideband** (SSB) operation. Bandwidth may be saved with the potential of doubling the channel capacity in a given band. Savings may also be made regarding transmitted power. One method used to produce SSB is by means of a balanced modulator to produce DSBSC and then filtering out either the lower sideband or upper sideband. SSB, as with DSBSC, requires a coherent detector at the receiver. Analysis of the effect of a phase offset ϕ in the receiver carrier signal used for detection[2] differs from that of DSBSC in that the amplitude of the demodulated signal is unaffected. Rather the demodulated signal is of the general form:

$$m(t) = V_\mathrm{m} \sin(\omega_\mathrm{m} t + \phi) \tag{4.38}$$

where we see that a phase shift results. It is equivalent to an additional time delay in transmission and is known as a **phase delay** ϕ which, if small, may be tolerated in audio systems. Its effect may be more critical in certain data communication and television systems.

4.2 Angle modulation

Instead of modulating amplitude, the phase angle of a carrier may be modulated by the message signal. Such a modulated wave, unlike AM where the intelligence is contained within the amplitude, has a constant amplitude. Angle modulation, especially FM, is therefore superior to AM in that noise effects,

---

[2] Connor, F.R., *Modulation*, 2nd edn, Edward Arnold, 1982, p. 125. ISBN 0-7131-3457-7.

which cause amplitude variation of the modulated signal, can largely be eliminated by limiting the signal prior to demodulation. As an example it is common knowledge that static interference is quite commonly experienced in AM radio reception. FM reception is generally immune to such amplitude-borne interference.

An angle modulated signal may be expressed:

$$g(t) = A \cos \theta(t) \, \text{V} \tag{4.39}$$

where the angle $\theta(t)$ is proportional to the amplitude of the modulating signal. As stated earlier, angle modulation can be subdivided into PM and FM.

4.2.1 Phase modulation

Suppose that the modulating signal $m(t)$ is a single sinusoid wave given by the expression:

$$m(t) = V_m \cos \omega_m t \tag{4.40}$$

Then a phase modulated wave may be expressed:

$$g(t) = A \cos[2\pi f_c t + \Delta\theta \cos(2\pi f_m t)] \tag{4.41}$$

where: $\cos 2\pi f_c t$ is the unmodulated carrier,
$\Delta\theta$ equals $k V_m$ where k is a constant of the modulator and has the unit of radian.

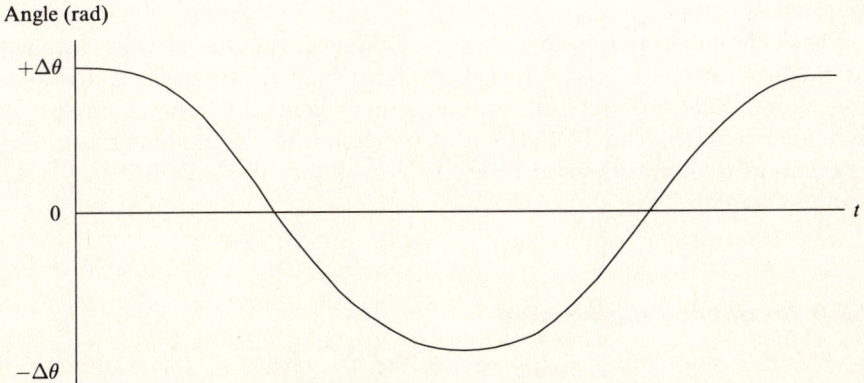

Angle (rad)

Figure 4.14   Phase modulation

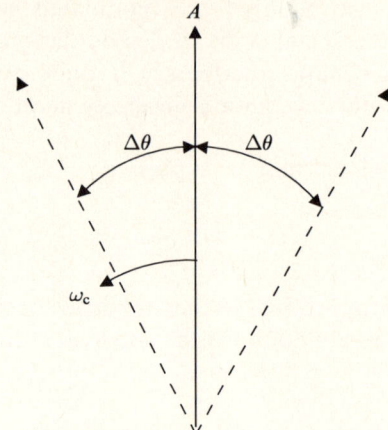

Figure 4.15   PM phasor diagram

The expression $\Delta\theta\cos(2\pi f_m t)$ can be shown graphically to base of time, Figure 4.14. It is apparent that the expression varies sinusoidally between the limits $\pm\Delta\theta$ radian. Therefore, by inspection of Eqn (4.41), a phase modulated wave has a centre, or mean frequency, f_c, the phase of which is sinusoidally modulated such that the peak *phase deviation* is $\pm\Delta\theta$ radian. Such a PM signal could be shown as a phasor diagram, Figure 4.15, where a phasor of amplitude A rotates with a mean angular velocity ω_c and oscillates between $\pm\Delta\theta$ about a mean value of zero.

In Figure 4.15, note that the rate at which the resultant phasor moves with respect to the unmodulated carrier is proportional to the frequency deviation about f_c which in turn is proportional to both f_m and V_m. The diagram also reveals that the modulated signal has an instantaneous frequency which varies sinusoidally about f_c.

Phase modulation is rarely used in analogue systems because frequency modulation can use relatively simple frequency discriminators for reception. Demodulation of phase modulation is generally more complex, one technique requiring an FM demodulator followed by an integrator. Phase modulation is, however, used extensively in digital transmission and is dealt with in Chapter 5.

4.2.2 Frequency modulation

In FM the modulating signal causes the frequency at the output of the modulator to vary about the unmodulated carrier frequency. As an example, the output frequency increases if the modulating signal goes positive, and vice

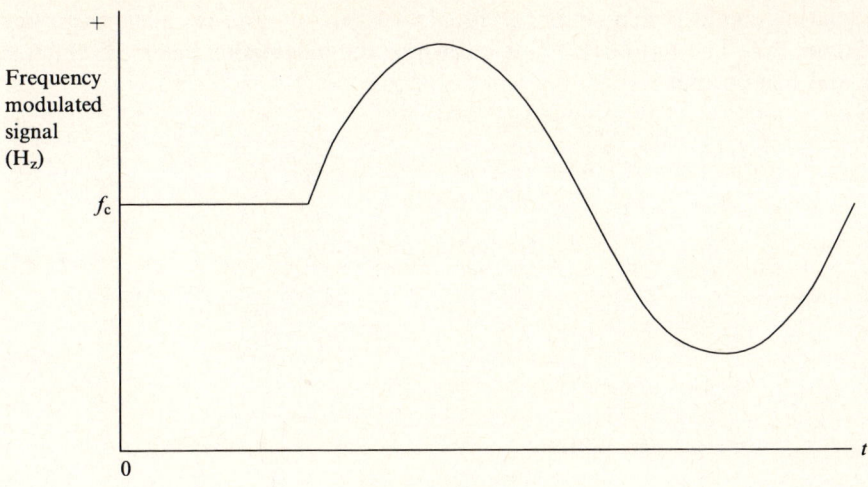

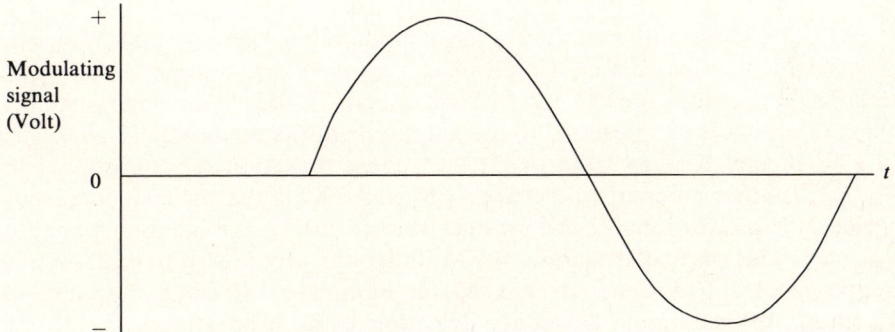

Figure 4.16   Frequency modulation

versa. The frequency *change*, rather than phase as in PM, is proportional to the amplitude of the modulating signal. An FM signal is shown in Figure 4.16.

An expression for an FM signal is given by:

$$g(t) = A \sin \left[2\pi f_c t + 2\pi k \int_0^t V_m \cos(2\pi f_m t)\, dt \right] \qquad (4.42)$$

$$= A \sin \left[2\pi f_c t + \frac{k V_m}{f_m} \sin(2\pi f_m t) \right] \qquad (4.43)$$

The term $k V_m$ must, from a dimensional analysis consideration, have the unit hertz and represents the maximum change in frequency produced by the modulating signal (when $\sin 2\pi f_\mathrm{m} t$ equals $+1$, or -1) and is called **frequency deviation**, ΔF. The term $kV_\mathrm{m}/f_\mathrm{m}$ is known as the **modulation index** M which is real and has no unit

$$M = \frac{kV_\mathrm{m}}{f_\mathrm{m}} \tag{4.44}$$

$$= \frac{\text{frequency deviation}}{f_\mathrm{m}} \tag{4.45}$$

$$= \frac{\Delta F}{f_\mathrm{m}} \tag{4.46}$$

Hence, $g(t)$ may also be written:

$$g(t) = A \sin(\omega_\mathrm{c} t + M \sin \omega_\mathrm{m} t) \tag{4.47}$$

PM and FM are both examples of angle modulation where the phase angle of the carrier is modulated and, if Eqns (4.41) and (4.47) are compared, it may be seen that expressions for PM and FM are similar. In PM, as already noted, the phase deviation is proportional to the amplitude of the modulating signal, V_m. In FM, it may be seen from (4.47) that phase deviation is proportional to $V_\mathrm{m}/f_\mathrm{m}$. Another comparison between FM and PM is that in PM, frequency deviation is proportional to the product $V_\mathrm{m} f_\mathrm{m}$. In FM, ΔF is proportional to V_m, only. This particular feature of FM illustrates why it is so named, i.e. the frequency of a FM signal is changed, or modulated, in direct response to $V_\mathrm{m} \sin \omega_\mathrm{m}$, the maximum frequency deviation being proportional to V_m, ω_m merely controlling the rate at which the FM signal frequency changes. A phasor diagram of FM is similar to that of PM. However, the peak phase deviation equals M. The phase is, as in PM, also modulated by the term $\sin \omega_\mathrm{m} t$.

The peak value of modulation index when frequency deviation and modulating frequency are at their maximum values is known as the **deviation ratio** and is a parameter of FM systems often defined in international regulations governing radio systems:

$$\text{deviation ratio} = \frac{\widehat{\Delta F}}{\widehat{f_\mathrm{m}}} \tag{4.48}$$

EXAMPLE 4.2
A domestic FM radio stereo broadcasting system has a system deviation of 75 kHz and maximum modulating frequency of 15 kHz. Determine the deviation ratio.

SOLUTION

$\widehat{\Delta F}$ equals $\pm 75\,\text{kHz}$

$\widehat{f_m}$ equals $15\,\text{kHz}$

$$\text{deviation ratio} = \frac{\widehat{\Delta F}}{\widehat{f_m}} \tag{4.49}$$

$$= \frac{75\,\text{kHz}}{15\,\text{kHz}} \tag{4.50}$$

Hence the deviation ratio is 5.

By means of a trigonometric identity the equation for an FM signal given earlier, Eqn (4.47) may be expanded and the spectrum of an FM signal, for a single modulating frequency, may be determined:

$$g(t) = A \sin \omega_c t \cos(M \sin \omega_m t) + A \cos \omega_c t \sin(M \sin \omega_m t) \tag{4.51}$$

Now, $\cos(M \sin \omega_m t)$ may be expressed in infinite series form:

$$\cos(M \sin \omega_m t) = 1 - \frac{M^2 \sin^2 \omega_m t}{2!} + \frac{M^4 \sin^4 \omega_m t}{4!} + \cdots \tag{4.52}$$

Now, by use of the following two identities:

$$\sin^2 \omega_m t = \tfrac{1}{2}(1 - \cos 2\omega_m t) \tag{4.53}$$

and

$$\sin^4 \omega_m t = \tfrac{3}{8} - \tfrac{1}{2} \cos 2\omega_m t + \tfrac{1}{8} \cos 4\omega_m t \tag{4.54}$$

given that higher-order powers of $\sin \omega_m t$ in Eqn (4.52) are all even, Eqns (4.53) and (4.54) may be used to substitute sin terms in Eqn (4.52) and, after simplifying, reveals that $\cos(M \sin \omega_m t)$ expands to the general form:

$$\cos(M \sin \omega_m t) = a_0 + a_2 \cos 2\omega_m t + a_4 \cos 4\omega_m t + \cdots \tag{4.55}$$

The terms a_0, a_2, a_4, etc., are merely coefficients that form a power series in M. In a similar manner an expression for $\sin(M \sin \omega_m t)$ may be developed, which is found to have the form:

$$\sin(M \sin \omega_m t) = a_1 \sin \omega_m t + a_3 \sin 3\omega_m t + a_5 \sin 5\omega_m t \ldots \tag{4.56}$$

Equations (4.55) and (4.56) could be substituted into Eqn (4.51). Clearly, the resultant equation would contain further products of sin and cos terms. These in turn generate the following spectral terms:

$$\omega_c \pm \omega_m, \ \omega_c \pm 2\omega_m, \ \omega_c \pm 3\omega_m, \ldots, \ \omega_c \pm \infty\omega_m$$

Hence, without recourse to full mathematical rigour, although you are referred to Stremler[3] for a complete analysis of the spectrum of an FM signal for a single modulating frequency, we may deduce that Eqn (4.51) takes the form:

$$A \sin \omega_c t \cos(M \sin \omega_m t) + A \cos \omega_c t \sin(M \sin \omega_m t)$$
$$= A[a_0 \sin_c t + a_1' \sin(\omega_c + \omega_m)t + a_2' \sin(\omega_c + 2\omega_m)t$$
$$+ a_3' \sin(\omega_c + 3\omega_m)t + \cdots + a_\infty' \sin(\omega_c + \infty\omega_m)t] \qquad (4.57)$$

Each of the coefficients in Eqn (4.57) is also a series in M and in fact takes the same form as Bessel functions of the first kind. As such the coefficients are normally written in the form $J_n(M)$ where n denotes the order and M is used to evaluate the particular coefficient sought. Equation (4.57) may be rewritten:

$$A \sin \omega_c t \cos(M \sin \omega_m t) + A \cos \omega_c t \sin(M \sin \omega_m t)$$
$$= A[J_0(M) \sin \omega_c t + J_1(M) \sin(\omega_c + \omega_m)t + J_2(M) \sin(\omega_c + 2\omega_m)t$$
$$+ J_3(M) \sin(\omega_c + 3\omega_m)t + \cdots + J_\infty(M) \sin(\omega_c + \infty\omega_m)t] \qquad (4.58)$$

A graph with curves of Bessel functions may be used to determine the values of $J_n(M)$ for each n and given M. Figure 4.17 illustrates such a graph.

The spectral content of the FM signal as represented by Eqn (4.58) has an infinite bandwidth. In practice an FM signal is band-limited. Inspection of the curves shown in Figure 4.17 reveals that for each value of n the coefficient $J_n(M)$, although oscillatory in nature, tends to zero as n approaches infinity. An FM signal which is produced by a single modulating sine wave, as in the above analysis, contains side frequencies spaced symmetrically, at harmonic intervals of ω_m, either side of the carrier to $\pm\infty$ Hz, Figure 4.18. Note that for certain values of M side frequency pairs, and indeed the carrier, may have zero amplitude.

System design will normally specify that bandwidth should encompass all significant sidebands for the value of deviation ratio in use where 'significant' is a subjective choice where the system is deemed acceptable to the user. For

---

[3] Stremler, F.G., *Introduction to Communications Systems*, Addison-Wesley, 1977, pp. 273–5. ISBN 0-201-07244-0.

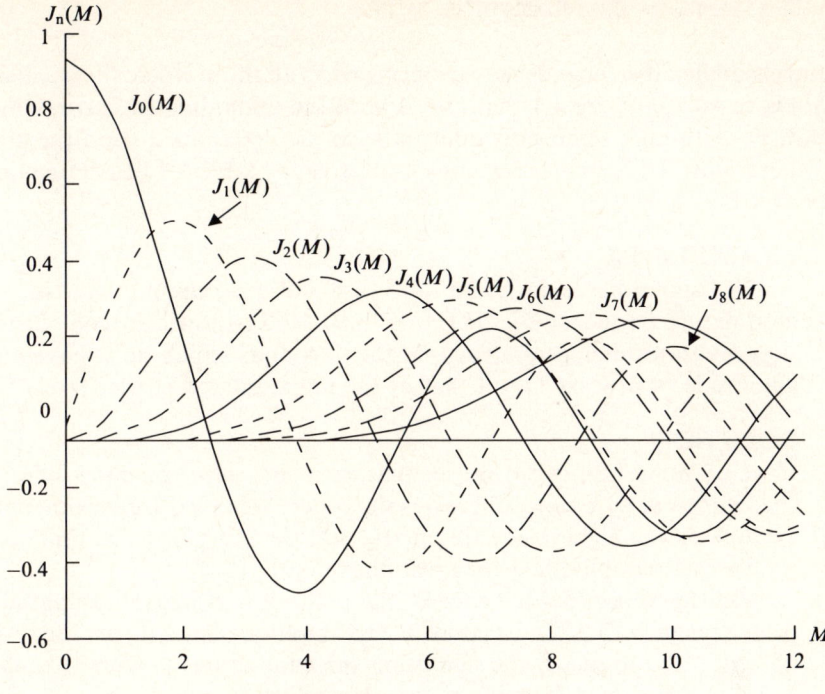

Figure 4.17 Bessel functions

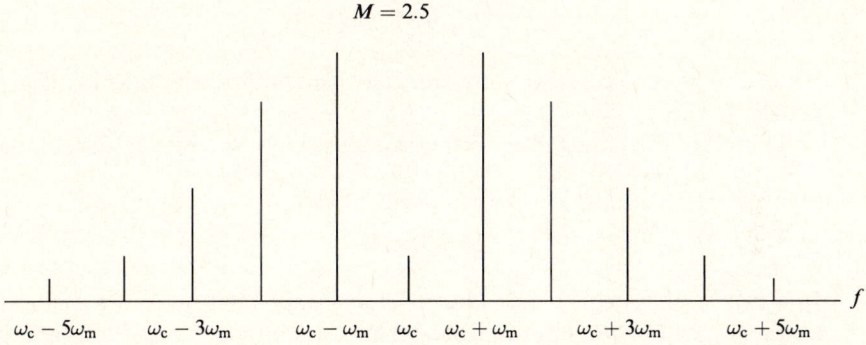

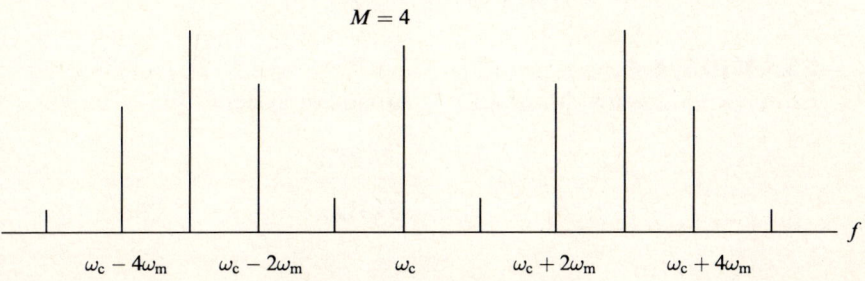

Figure 4.18 FM spectrum

example significant sidebands may be regarded as all those where the coefficient $J_n(M)$ is equal to, or greater than, say 1% of the unmodulated carrier amplitude $J_0(0)$. Although significant sidebands can be determined from the graph shown in Figure 4.17, greater accuracy is afforded by tables of Bessel functions, Appendix D.

EXAMPLE 4.3

An FM signal has a deviation ratio of 5 and a maximum value of modulating frequency of 15 kHz. Estimate the bandwidth required to support transmission of all significant sidebands which must have an amplitude of at least 1% of that of the unmodulated carrier level.

SOLUTION

If we examine Eqn (4.58) we see that each sine term generates side frequencies at $\omega_c \pm n\omega_m$, where n is the order. What we are seeking in this example is the highest side frequency, and therefore order n, that produces an amplitude $J_n(M)$ just above 0.01.

With reference to Appendix D, the unmodulated carrier amplitude occurs when both M and n equal 0. The amplitude has a normalised value of 1. In order to satisfy the significant sideband criterion with M equal to 5 the modulus of $J_n(5)$ must satisfy the following criteria:

$$J_n(5) > 0.01$$

and

$$J_{n+1}(5) < 0.01$$

From the Bessel tables we may see that this condition is satisfied if n equals 8.

Therefore the bandwidth must encompass eight sideband pairs, that is:

$$W = 2(8 \times 15\,\text{kHz})$$

$$= 240\,\text{kHz}$$

In a practical system, bandwidth is largely dependent upon the highest frequency within the information signal. Carson has produced a 'rule of thumb' for bandwidth, known as Carson's rule:

$$\text{Bandwidth} = 2(\widehat{\Delta F} + \widehat{f_m})$$

EXAMPLE 4.4

Estimate the bandwidth of an FM broadcast system, if:

$$\widehat{\Delta F} = \pm 75\,\text{kHz}$$

$$\widehat{f_m} = 15\,\text{kHz}$$

SOLUTION

$$B = 2(\widehat{\Delta F} + \widehat{f_m})$$
$$= 2(75 + 15)\,\text{kHz}$$
$$= 180\,\text{kHz}$$

Both this example and Example 4.3 are based upon the standard FM radio stereo broadcasts first seen in Example 4.2. In comparing the bandwidth required using Carson's rule with that obtained using Bessel tables, results differ slightly. In practice a bandwidth of 200 kHz is typical which is a compromise between the two figures.

4.2.3 Narrowband FM

The description of FM in this chapter so far has been for the general case. Where $M > 1$ many side frequencies are produced and such a signal is known as a **wideband FM** signal. If $M < 1$, **narrowband FM** is produced. It follows that narrowband FM results in a much smaller bandwidth than is the case for wideband FM. In addition, because the signal is contained within a relatively narrow bandwidth, the effect of noise is more pronounced. Hence narrowband FM requires a larger S/N ratio than is the case for wideband FM.

An expression for an FM signal was shown in Eqn (4.51) as:

$$g(t) = A \sin \omega_c t \cos(M \sin \omega_m t) + A \cos \omega_c t \sin(M \sin \omega_m t) \qquad (4.59)$$

If $M \ll 1$, then:

$$\cos(M \sin \omega_m t) \to 1$$

and, since $\sin x$ approaches x as x approaches zero:

$$\sin(M \sin \omega_m t) \to M \sin \omega_m t$$

Therefore a narrowband FM signal approximates to:

$$g(t) \approx A[\sin \omega_c t + \cos \omega_c t(M \sin \omega_m t)] \qquad (4.60)$$

$$= a\left[\sin \omega_c t + \frac{M}{2}\sin(\omega_m - \omega_c)t + \frac{M}{2}\sin(\omega_m + \omega_c)t\right] \qquad (4.61)$$

and, by rearranging, we may write:

$$g(t) = a\left[\sin \omega_c t - \frac{M}{2}\sin(\omega_c - \omega_m)t + \frac{M}{2}\sin(\omega_c + \omega_m)t\right] \qquad (4.62)$$

The above equation is similar to that of a full AM signal shown earlier in Eqn (4.5). In comparing narrowband FM with AM, we see that its side frequencies are of *opposite* amplitude and that both forms of modulation have the same bandwidth, namely twice that of the modulating signal. A phasor diagram to illustrate narrowband FM is shown in Figure 4.19 where it may be seen that, unlike AM, anti-phase side frequencies give rise to both amplitude *and* phase modulation of the carrier. Both of these features may be used for reception purposes.

Some FM communication systems are inherently narrowband, most notably analogue mobile radio systems such as CB radio and public mobile radio systems. Such speech-based communication employs FM and typically operates with an audio bandwidth of 3 kHz. Owing to the scarcity of available spectrum, systems are required to operate on a narrow band basis with a typical channel spacing of the order of 20 kHz.

Narrow band FM is a compromise between the minimal bandwidth requirements of AM and the superior noise immunity (to amplitude variation) of wideband FM. It also has application in the generation and reception of wideband FM and PM signals.

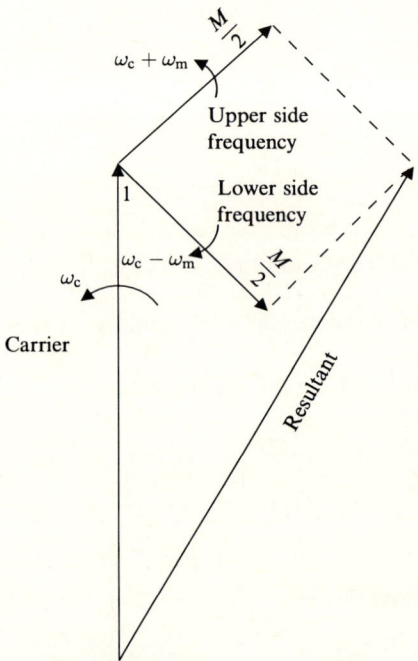

Figure 4.19   Narrowband FM phasor diagram

4.2.4 FM generation and demodulation

Detail regarding the variety of techniques used to produce and demodulate FM is beyond the scope of this text. Suffice to say here that FM modulation is produced either directly or indirectly. The **direct** method arranges that the message signal directly modulates the frequency of an oscillator circuit. Commonly a varactor (variable reactance) diode is made use of. When a p–n junction diode is reverse biased, the depletion layer exhibits the property of a dielectric which forms capacitance. A modulating signal may be applied to vary the degree of reverse bias voltage which in turn creates a variable capacitance. If the varactor diode is placed in the frequency-determining element of an oscillator, FM may be produced.

The direct method for wideband modulation, with relatively large frequency deviation, cannot be achieved using crystal oscillators. In consequence wideband modulation requires great effort to be made in the design of the carrier frequency oscillator to satisfy the close frequency tolerances required for radio transmission. The stability benefits of a crystal may be realised by means of **indirect** modulation. This is typically achieved using the 'Armstrong' method whereby phase modulation, by means of suitable pre-processing of the modulating signal, is produced using a crystal oscillator. As we saw earlier, PM exhibits an element of FM, which is narrowband in nature. In order to achieve a wideband FM signal, with sufficient frequency deviation and at the correct carrier frequency, a number of frequency multiplier and mixer stages are usually employed.

FM demodulators are commonly termed discriminators which ideally produce an output amplitude that is linearly proportional to the frequency of the received FM signal. The most popular type of discriminator is a phase locked loop where a voltage controlled oscillator (VCO) tracks the instantaneous frequency of the FM modulated signal. The control voltage of the VCO is in effect the demodulated signal. In the case of FM demodulation a limiter immediately precedes the discriminator to remove any amplitude variation, which is assumed to be noise, in the FM signal, thereby enhancing the S/N ratio at the output of the receiver.

4.3 Frequency division multiplexing

Frequency division multiplexing (FDM) is the process of combining a number of signal sources, or channels, into a single composite signal. This is achieved by modulating each signal onto a unique carrier frequency carefully chosen so that the spectra of each modulated signal do not overlap. In this way a transmission bandwidth is divided up into a number of separate frequency bands, each of which accommodates one signal: hence the term 'frequency division'.

The classic example of FDM is the carrier frequency line communication system (commonly referred to as a **carrier system**). Carrier systems are one of the earliest multi-channel systems developed and enable as many as 60 telephone channels to be transmitted over conventional copper wire. Signals are split into the two directions of signal transmission (Go and Return) and a separate pair assigned for each. This enables the use of amplifiers, which are inherently unidirectional, to support long-distance operation for trunk telephony links between cities.

Figure 4.20 illustrates how 12 channels are formed into a basic **FDM group** occupying the frequency band 12–60 kHz. Each channel is modulated and filtered to produce an SSB signal. Channels are allocated 4 kHz of bandwidth. Telephone signals are constrained to a frequency range of 300–3400 Hz and hence each channel has a **guard band** in practice to prevent adjacent channel interference.

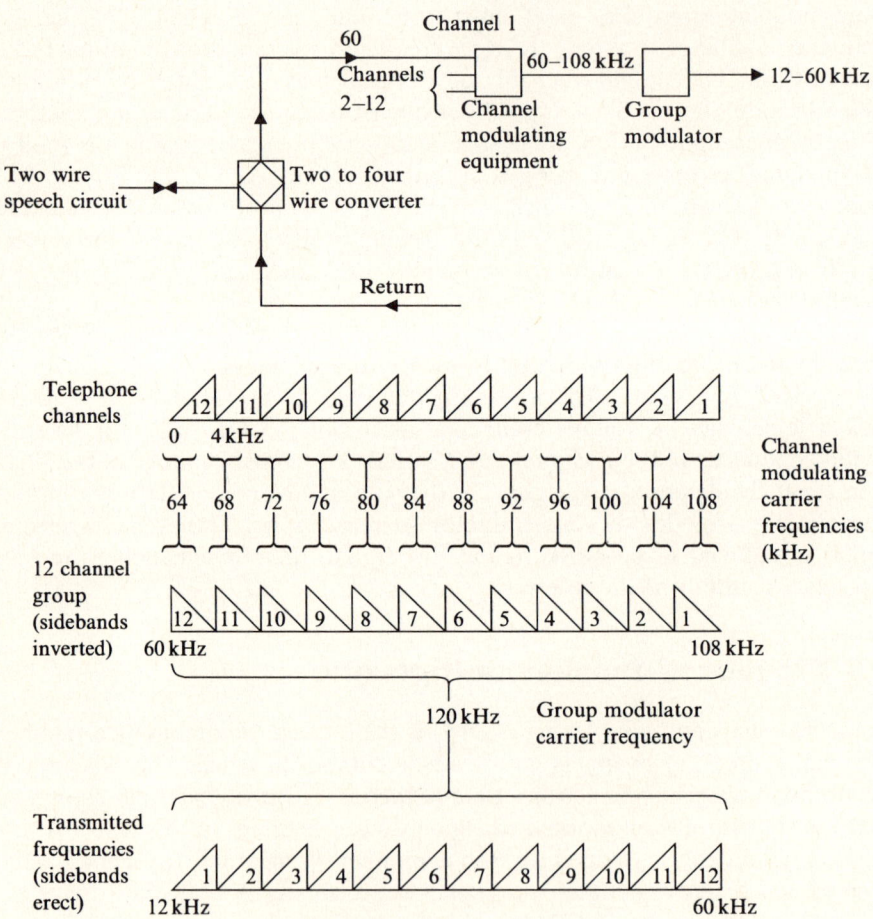

Figure 4.20   FDM carrier system

Channel filters are employed after each modulator, again to prevent co-channel interference, before signals are combined and passed to a **group modulator**. Crystal filters are chosen because they exhibit a very sharp cut-off. To obtain sharper responses, a group is initially formed in the frequency range 60–108 kHz. Channel filters are so arranged that the lower sideband of each channel is selected. The lower sideband is said to be **inverted** because the highest frequencies of the signal prior to modulation appear as the lowest frequencies after modulation, and vice versa. All of the frequencies are then changed, or **translated**, by the group modulator by modulating with a carrier frequency of 120 kHz. Selection of the lower sideband frequencies results in the desired frequency band of 12–60 kHz. The sideband of each individual channel within the group is now said to be **erect**; that is, direct correspondence between high and low frequencies before and after modulation.

Although not shown, complementary equipment is used for the return direction signals comprising a group demodulator to convert signals back into the band 60–108 kHz followed by channel demodulating equipment. The system was exploited much further. Five 12-channel groups could be further modulated to form an FDM signal comprising 60 channels, still capable of transmission over conventional cables. Further multiplexing led eventually to systems that operated at megahertz but that now required lower loss cable, which was afforded by coaxial transmission media.

Many carrier systems have been superseded by digital multiplexing schemes which are discussed in Chapter 5. However, the above illustrative example is useful to demonstrate the concept of FDM. It is important in communications because it is the way that many multi-channel systems make use of radio transmission to ensure channel spectra do not overlap and interfere. Examples include national television broadcast distribution, cellular radio systems used for mobile telephones and terrestrial and satellite microwave radio links.

4.4 Comparison of modulation methods

The two main parameters to consider in comparing different types of modulation are bandwidth and S/N ratio. The S/N ratio is considered both at a receiver input for satisfactory operation, and also at the receiver output. Since we have already discussed bandwidth requirements of modulated signals, we shall examine their S/N performances and conclude with a summary.

4.4.1 Threshold effect

Before comparing modulation methods, consider the effect of noise upon a full AM signal where noise may be represented:

$$n(t)\cos[\omega_c t + \phi(t)]$$

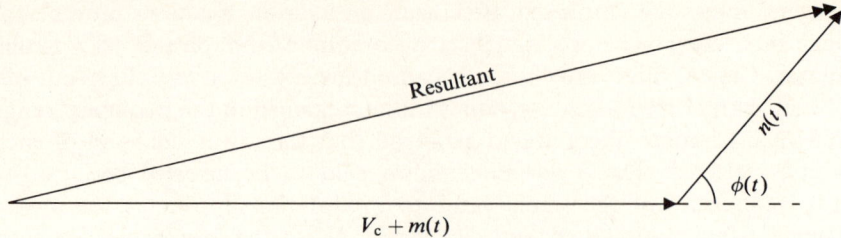

Figure 4.21 AM signal in presence of noise

The equation for an AM signal shown in Eqn (4.4) may alternatively be expressed:

$$[V_c t + m(t)] \sin \omega_c t \qquad (4.63)$$

where, as earlier:

$$m(t) = V_m \sin \omega_m t \qquad (4.64)$$

Hence, the signal received at the input to the demodulator may be expressed:

$$g(t) = [V_c + m(t)] \cos \omega_c t + n(t) \cos[\omega_c t + \phi(t)] \qquad (4.65)$$

A static phasor diagram representation of Eqn (4.65) is shown in Figure 4.21.

When $[V_c + m(t)] \gg n(t)$, S/N may be regarded as high and Betts[4] shows that an expression for the output of an envelope detector, V_{out}, may be approximated:

$$V_{out} \approx \underbrace{[\,V_c + m(t)\,]}_{\text{signal}} + \underbrace{n(t) \cos \phi(t)}_{\text{noise}} \qquad (4.66)$$

Now consider low S/N conditions when $n(t) \gg [V_c + m(t)]$. Hence, from Eqn (4.65):

$$v_{out} \approx \underbrace{n(t) + V_c \cos \phi(t)}_{\text{noise}} + \underbrace{m(t) \cos \phi(t)}_{\text{signal} \times \text{noise}} \qquad (4.67)$$

In comparing Eqns (4.66) and (4.67) we see that the detector output consists of noise components that are additive when S/N is high or multiplicative when S/N is low. Clearly the effect of multiplying the signal by the noise component $\cos \phi(t)$ will have far greater influence upon the output than noise that is

---

[4] Betts, J.A., *Signal Processing, Modulation and Noise*, The English Universities Press Ltd, 1970, pp. 98–99. ISBN 0-340-09895-3.

merely additive, in which case it is also relatively weak. The transition between an additive effect and a multiplicative effect occurs at a certain threshold level. The exact value of S/N at which **threshold effect** occurs is imprecise and may only be determined empirically. The multiplicative property of noise under poor S/N conditions is characteristic of all non-coherent detectors, namely full AM and angle demodulators.

4.4.2 Comparison of demodulation processes

Figure 4.22 shows a generalised block diagram of a receiver. A pre-detection filter, typically an intermediate frequency (IF) stage in a radio receiver, limits the bandwidth of the received signal to that of the modulated signal to prevent any out of band noise contributing to the noise power at the demodulator output. $(S/N)_{in}$ is simply the S/N ratio prior to demodulation, or detection. Dependent upon the nature of the demodulation process a post-detection filter may, or may not, be required to eliminate any frequency components that may exist above that of the bandwidth of the recovered message signal. $(S/N)_{out}$ is the recovered message signal and associated noise.

Baseband transmission

We shall now determine $(S/N)_{out}$ for a variety of demodulation techniques. We shall assume that the message signal has a bandwidth of W Hertz and that the transmission system only introduces Gaussian noise with a double-sided PSD of amplitude $\eta/2$, as shown earlier in Figure 3.4(b).

Firstly we shall consider $(S/N)_{out}$ for baseband transmission to which modulated systems may later be compared. The system bandwidth equals that of the message signal W. Hence we may state:

$$(S/N)_{in} = \frac{S_R}{\frac{\eta}{2} \cdot 2W} \qquad (4.68)$$

$$= \frac{S_R}{\eta W} \qquad (4.69)$$

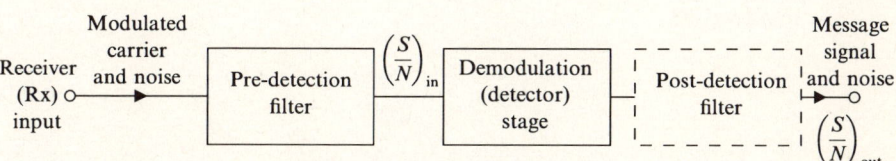

Figure 4.22 Generalised receiver

where S_R is the signal power at the receiver input. In the case of baseband transmission there is no demodulation stage and we may write:

$$(S/N)_{out} = (S/N)_{in} \qquad (4.70)$$

$$\therefore \quad (S/N)_{out} = \frac{S_R}{\eta W} \qquad (4.71)$$

$(S/N)_{out}$ is plotted as a function of $(S/N)_{in}$ in Figure 4.23.

Full AM: coherent and envelope detection

Full AM has a system bandwidth of $2W$. In the case of coherent detection and envelope detection above threshold, that is providing $(S/N)_{in}$ is at least 10 dB, we may, with reference to Figure 4.22, write:

$$(S/N)_{in} = \frac{S_R}{2\eta W} \qquad (4.72)$$

and

$$(S/N)_{out} = \frac{S_R}{\eta W} \qquad (4.73)$$

Here we see that (S/N) after detection is double that received; that is, there is a 3 dB improvement in S/N in the detection process. The value of $(S/N)_{out}$ is the same as in the case of baseband transmission. However, the signal input to the demodulation process contains a carrier which appears at the output of the demodulator as a dc component corresponding to the unmodulated carrier's peak amplitude. Since this is not related to the original information signal, its power must be ignored. In consequence when measuring S_{out}, dc power should be ignored, hence $(S/N)_{out}$ in practice is effectively 7–10 dB worse than that indicated in Eqn (4.73). A modified value of $(S/N)_{out}$ for full AM, reflecting the above, is shown in Figure 4.23 which is adjusted by 10 dB compared with baseband transmission.

DSBSC

Analysis of DSBSC is similar to full AM, both systems having the same transmission bandwidth:

$$(S/N_{in}) = \frac{S_R}{2\eta W} \qquad (4.74)$$

$$(S/N)_{out} = \frac{S_R}{\eta W} \qquad (4.75)$$

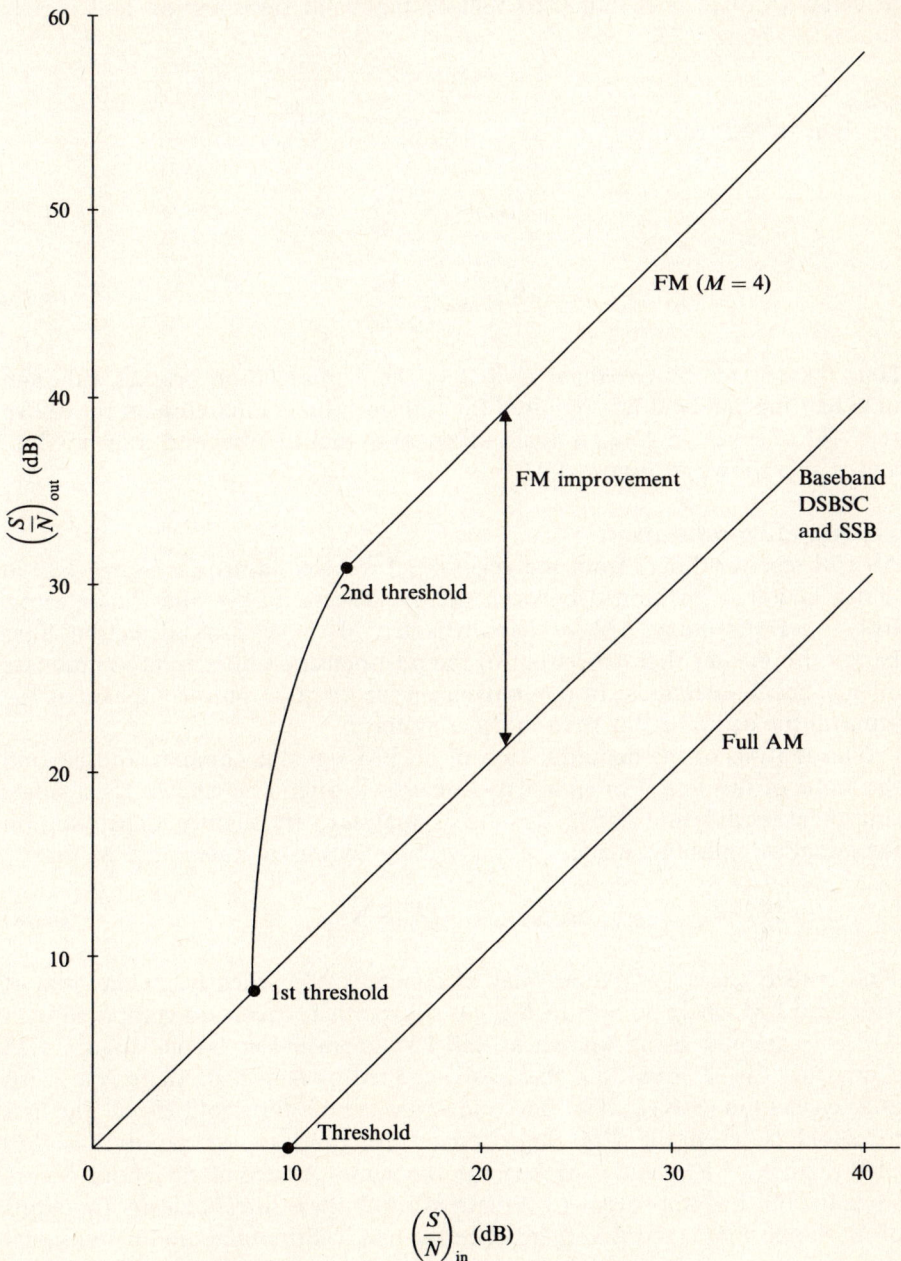

Figure 4.23 Comparison of modulation techniques

Here, as with full AM, we see a 3 dB improvement in S/N ratio through the detector. $(S/N)_{out}$ is identical to that of baseband transmission and is also shown in Figure 4.23.

SSB
Analysis of SSB reveals:

$$(S/N)_{in} = \frac{S_R}{\eta W} \tag{4.76}$$

$$(S/N)_{out} = \frac{S_R}{\eta W} \tag{4.77}$$

Here there is no improvement in S/N in the demodulation process although only half the bandwidth is required for transmission compared with the above two AM schemes. $(S/N)_{out}$ is again identical to that of baseband transmission, and is also shown in Figure 4.23.

Frequency modulation
An FM receiver differs from the generalised receiver shown in Figure 4.22 in that a limiter is positioned between the pre-detection filter and the detector. FM and PM, unlike AM or narrowband FM, have a pre-detection filter bandwidth greater than twice that of the post-detection filter to accommodate all significant sidebands. In consequence a larger noise power appears at the input to the demodulator than in other systems.

The analysis of the demodulation of an FM signal is complex and beyond the scope of this text. For simplicity, analysis is often only shown for a single sinusoidal modulating signal. Results of analyses vary slightly, depending on the initial assumptions made. An analysis by Schwartz[5] states for FM that:

$$(S/N)_{out} = 3M^2(S/N)_{in} \tag{4.78}$$

That is $(S/N)_{out}$ is $3M^2$ times that of coherent AM. Clearly, in the case of wideband FM where $M > 1$, $(S/N)_{out}$ is substantially increased compared with AM, the amount being known as the **FM improvement factor**. Figure 4.23 shows that for M equal to 4 then, as suggested by Schwartz, there is a 13 dB improvement in $(S/N)_{out}$. FM does, however, exhibit threshold effect. The first threshold is when the FM improvement collapses to the point that S/N performance of FM and AM become the same. A second threshold occurs when the full FM improvement is reached. Both these thresholds are functions of M. Suppose M is reduced then, since the bandwidth of the modulated signal

---

[5] Schwartz, M., *Information Transmission, Modulation and Noise*, 4th edn, McGraw-Hill, pp. 510–17. ISBN 0-07-100931-0.

and hence noise power is less, the received signal can be reduced more before threshold effect is reached. That is FM improvement may be achieved with a lower $(S/N)_{in}$ as M is reduced.

Carson's rule may be rewritten:

$$B_T = 2f_m(1 + M) \tag{4.79}$$

where the bandwidth of the transmitted FM signal, or **transmission bandwidth**, is denoted by B_T. In the case of wideband FM, where M is much greater than 1, Eqn (4.79) may be approximated:

$$B_T \cong 2Mf_m \tag{4.80}$$

Rearranging Eqn (4.80) to make M the subject of the formula and substituting into Eqn (4.78) yields:

$$(S/N)_{out} \cong 3 \left(\frac{B_T}{2f_m} \right)^2 (S/N)_{in} \tag{4.81}$$

Equation (4.81) indicates that the FM improvement in S/N ratio is approximately proportional to the square of the transmission bandwidth. FM is an example of a **bandwidth expansion** system where bandwidth is traded-off to obtain superior signal to noise performance.

Equation (4.78) suggests that if M is increased indefinitely then FM may continue to shown an improvement in S/N ratio at the receiver output. If the transmitted power remains constant as M (and hence bandwidth) increases, the amplitude of S_R decreases. Since the noise power $\eta/2$ at the input is constant over all frequencies, if the level of S_R falls, so too will $(S/N)_{in}$ at the input to the demodulator. Eventually $(S/N)_{in}$ will become so low that threshold effect will take over, leading thereafter to a dramatic reduction in $(S/N)_{out}$.

The above discussion for FM is based upon wideband FM where $M \geqslant 0.6$ is the usual assumption. If M equals 0.6 then, from Eqn (4.78) we see that the FM improvement for narrowband operation is equal to, or less than, 0.3 dB compared with coherent detection; that is, the improvement is negligible which may be expected because, as seen earlier, the bandwidth of a narrowband FM signal is similar to that for AM.

4.4.3 Comparison of modulation techniques

Figure 4.23 and Table 4.1 provide useful comparisons of the various methods of modulation. The actual modulation method used in a particular application will depend upon the criteria governing the system. For instance if bandwidth is at a premium then clearly SSB would appear as the most attractive. Full AM

Table 4.1 Comparison of modulation techniques

Type of modulation	Band-width	Threshold effect	S/N improvement (cf. DSBSC)	Reception	Comment
Full AM	2W	Yes	−7 to −10 dB	Envelope detector or square law	Simple and cheap. Carrier-transmission wastes power. Use − broadcast.
DSBSC	2W	No	N/A	Coherent	Requires synchronous detection. SSB preferred due to reduced bandwidth requirement.
SSB	W	No	No	Coherent	Synchronous detection. Spectrally efficient. Widely used for telephony.
FM	≫W	Yes FM improvement generally assumed >10 dB	Yes	Frequency discriminator	Bandwidth expansion system. Improves S/N. Use − terrestrial, satellite microwave communication.
PM	≫W	Yes	Yes	Non-linear	Similar performance to FM. More complex reception.

W = baseband signal bandwidth.

on the other hand offers what is probably the cheapest receivers if an envelope detector is used. This feature is especially attractive in high-volume applications such as home entertainment receivers used for television and radio broadcasting. Clearly for best quality operation FM is best, but only at the expense of considerably increased bandwidth. It is nevertheless interesting to note that in poor S/N environments FM improvement diminishes and there may be no advantage in using it compared with AM.

4.5 Summary

Modulation is the process of impressing information upon another waveform of higher frequency called a carrier. Modulation is employed to convert signals of lower frequency into sufficiently high frequency for transmission over radio or optical fibre channels. A carrier may have either its amplitude (AM) or angular velocity modulated although the latter may be further categorised as either PM or FM. A key parameter of full AM is modulation depth which, apart from governing the shape of the modulated envelope, indicates the degree of noise immunity and power distribution in carrier and side frequency components. AM may be produced using either non-linear or linear techniques. Non-linear production makes use of deliberate operation over the non-linear characteristic of an active device and gives rise to harmonic distortion. Linear operation avoids such operation, leading to a modulated signal with less distortion. A diode detector is a cheap, and efficient, way of demodulating a full AM signal. Components must be carefully chosen to minimise negative peak clipping, which is a form of distortion in the recovered signal. DSBSC may be produced by means of a balanced modulator where carrier and modulating signal are directly multiplied. A feature of DSBSC is its absence of carrier components, offering the advantage of reduced transmitter power requirements. However, a coherent detector must be used which means that the original carrier must be recovered at the receiver in some way before demodulation can be performed.

Two common balanced modulator circuits are the Cowan and ring modulators. SSB operation, which also requires coherent detection, may be used where available bandwidth is at a premium. Mathematical analysis of the spectrum of an FM signal reveals it to be infinite. Bessel functions may be used, in conjunction with the modulation index, to determine the number of significant sidebands required and the corresponding bandwidth necessary. Carson's rule may alternatively be used to approximate the bandwidth of an FM signal. Narrowband FM is a compromise between AM and FM which seeks to obtain the superior noise performance of FM with smaller bandwidth approaching that of AM.

FDM is a common technique of multi-channel operation whereby two, or more, channels may be transmitted over a single link. FDM may be employed in cable systems, whether metallic or optical fibre, to increase the number of signals transmitted and so improve utilisation and economic efficiency. Radio systems also use FDM as a way of staggering channels spectrally to avoid interference. Threshold effect indicates that providing signal to noise ratio is high the noise has little additional effect at a detector output. Where this is not the case, the signal to noise ratio at the detector output deteriorates rapidly with a reduction in signal to noise.

The various forms of modulation are compared in the presence of noise, in conjunction with equivalent baseband operation. There is a marked increase in bandwidth required by an FM signal compared with AM. Full AM requires a greater transmitted power for a given signal to noise ratio at the detector output compared with other forms of modulation. This is partly due to the power 'wasted' in the carrier. FM shows an improved signal to noise ratio which may be increased further with modulation index. However, this in turn requires a corresponding increase in required bandwidth. FM is for this reason known as the bandwidth expansion system.

Exercises

4.1   Explain why the terms 'amplitude modulation' and 'angle modulation' are so named. What is the principal difference between the two forms of modulation?

4.2   Sketch carefully to scale, using squared paper, an AM waveform with a carrier of amplitude 4 V and a modulation depth of 0.35.

4.3   An AM signal has a modulation depth of 120%. Sketch the modulated waveform, illustrating what distortion results. Explain what would happen if such an AM signal was applied to a diode detector circuit.

4.4   An AM wave is represented by the expression:

$$g(t) = 10(1 - \cos 6280t) \sin(6\pi 10^6 t - \pi/6) \, \text{volt}$$

Determine:

(a)   the modulation depth;
(b)   the modulating frequency;
(c)   the period of the carrier;
(d)   the peak instantaneous voltage;
(e)   the bandwidth of the modulated signal.

4.5   The envelope of a full AM wave has a maximum amplitude of 4.5 V and a minimum amplitude of 2.3 V. Determine the modulation depth.

4.6   An AM wave is represented by the expression:

$$g(t) = 24(1 + 0.5 \cos 3140t) \sin 2\pi 10^5 t \, \text{volt}$$

If the modulated signal is applied to an amplifier of input resistance $600 \, \Omega$, calculate the power dissipated.

4.7   A full AM signal is transmitted with total power equal to 10 kW. If the modulation depth is 25%, calculate sideband and carrier powers.

4.8   A carrier frequency of 6 MHz is amplitude modulated by a signal with a frequency range from 300 to 3400 Hz. Sketch a diagram to illustrate the spectrum of the modulated signal.

4.9   In recovery of an AM signal, explain what happens to the carrier component's power at the receiver.

4.10 A DSBSC signal can be produced by adding a modulating signal and carrier together and applying to a non-linear circuit. The output of such an arrangement includes a DSBSC signal.

If the addition of a modulating signal and carrier is represented thus:

$$g(t) = V_m \sin \omega_m t + V_c \sin \omega_c t$$

and a non-linearity is defined as:

$$h(t) = k + aV_{in} + b(V_{in})^2$$

show mathematically that a component of the output voltage $k(t)$ is produced that is equivalent to a DSBSC signal.

4.11 An FM system has a maximum frequency deviation of 100 kHz and a modulating frequency of 15 kHz.

(a) Calculate the deviation ratio.
(b) Use Carson's rule to estimate a suitable bandwidth for this signal.
(c) Determine a suitable bandwidth using Bessel functions.

4.12 Explain 'threshold effect'. Your explanation should include appropriate sketch/es.

4.13 Compare full AM, DSBSC, SSB and FM with respect to bandwidth and S/N.

4.14 FM is sometimes referred to as a 'bandwidth extension system'. Explain what this might mean regarding its S/N improvement over that of AM systems.

Bibliography

Betts, J.A., *Signal Processing, Modulation and Noise*, The English Universities Press Ltd, 1970. ISBN 0-340-09895-3.

Connor, F.R., *Modulation*, 2nd edn, Edward Arnold, 1982. ISBN 0-7131-3457-7.

Schwartz, M., *Information Transmission, Modulation and Noise*, 4th edn, McGraw-Hill, 1990. ISBN 0-07-100931-0.

Stremler, F.G., *Introduction to Communications Systems*, 3rd edn., Addison-Wesley, 1990. ISBN 0-201-51651-0.

Zeimer, R.E. and Tranter, W.H., *Principles of Communications*, 4th edn, John Wiley & Sons, 1995. ISBN 0-471-12496-6.

CHAPTER 5

Digital communication

Aims and objectives

The effect of transmitting a digital signal over a practical channel that is band-limited and how inter-symbol interference occurs are considered. The use of an eye diagram is demonstrated in examining the degree of inter-symbol interference. Equalisation, in particular the use of a raised cosine response, is explained as a means of reducing the effect of inter-symbol interference. The effect of noise upon bit error rate at a receiver is considered. Channel capacity is defined which indicates that in any practical channel, there is a limit to the rate at which data may be transmitted. The concept of time division multiplexing is presented to illustrate how multiple digital signals may share a common transmission path. A pulse code modulation system is described. Signal to quantisation noise ratio is defined. The bandwidth of a pulse code modulation system is estimated and the use of compression explained. Digital modulation by means of ASK, FSK, PSK and QAM is explained. The effect of noise for each type of modulation is considered and the bit error rate compared against signal to noise ratio of the channel. The chapter concludes by comparing the performance of each form of modulation.

5.1 Baseband transmission

Signals may be transmitted over metallic conductors, optical fibres or radio systems. Although all three types of channel may make use of modulation for transmission, baseband transmission is only possible over metallic conductors. In this section we shall look at baseband transmission techniques in detail.

5.1.1 Band-limited channels

A metallic transmission line may be represented by an equivalent electrical circuit[1] which includes series inductance and shunt capacitance. In conse-

---

[1] Dunlop, J. and Smith, D.G., *Telecommunications Engineering*, 3rd edn, Chapman & Hall, 1994, p. 174. ISBN 0-412-56270-7.

quence transmission lines exhibit a low-pass filtering effect which gives rise to both attenuation and phase distortion.

Theoretically digital signals have an infinite spectrum. The low-pass filtering effect of a transmission line, or indeed any practical channel, does not automatically mean that perfect recovery of a digital signal at a receiver is not possible. Signalling rate R has the unit baud and equals the reciprocal of the duration of the shortest signalling element. Perfect recovery is possible providing that the signalling rate does not exceed twice the available bandwidth. This rate is known as the **Nyquist rate**[2]:

$$R = 1/T \leqslant 2W \tag{5.1}$$

where T is the duration of one signalling element, or **symbol**. Alternatively the Nyquist limit may be regarded as transmission of a maximum of two symbols per hertz of bandwidth. There is a similarity with sampling in Chapter 2 where the number of samples that may be transmitted per hertz of bandwidth is also two. Too high a sampling rate results in aliasing. So too if digital signals are transmitted at a rate in excess of the Nyquist rate aliasing occurs at the receiver.

Analysis of the practical effect of the low-pass filtering effect of a transmission line upon a train of pulses requires exact knowledge of a transmission line's particular transfer function. Consider the effect of a perfect filter upon a pulse train where the line's frequency response is assumed to be ideal as shown in Figure 5.1(a). We shall assume a signalling rate equal to exactly twice the filter's bandwidth. That is:

$$1/T = 2W \tag{5.2}$$

or:

$$T = \tfrac{1}{2}W \tag{5.3}$$

The effect of such filtering may be found by convolving the time domain response of the filter with that of the digital signal. From the Fourier transform tables in Appendix C we may determine the time domain response of the filter by using the inverse Fourier transform of a rectangular function in conjunction with frequency scaling by a factor W:

$$H(f/2W) \rightleftharpoons \tfrac{1}{2}h(t2W) \tag{5.4}$$

[2] Nyquist, H., 'Certain topics in telegraph transmission theory', *AIEE Trans.*, **47**, 617–44, 1928.

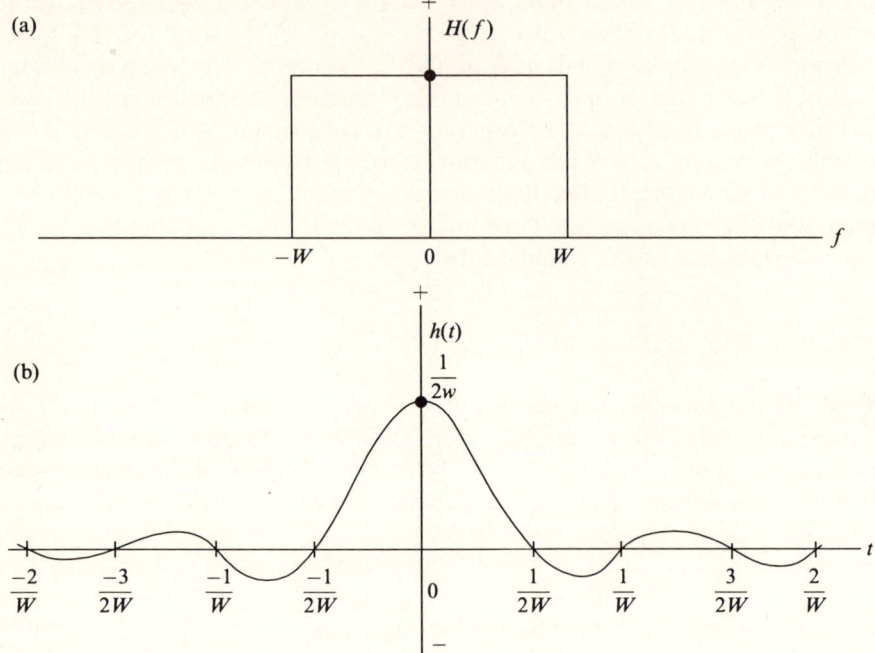

Figure 5.1   Time domain response of ideal low-pass filter (a) and with inverse Fourier transform (b)

Therefore:

$$\text{rect}(f/2W) \rightleftharpoons \text{sinc}(2Wt) \tag{5.5}$$

Hence $h(t)$ for an ideal low-pass filter is as shown in Figure 5.1(b).

Determination of the effect of convolving a digital signal consisting of pulses with the above sinc pulse is not trivial. For simplicity we shall assume that each pulse is a delta or impulse function and that the signal is bipolar. Such a signal is shown in Figure 5.2(a).

Convolution of a single impulse function with the sinc function response of the filter, Figure 5.1(b), produces an identical sinc function, but centred upon the instant in time of the particular impulse function. Superposition may be brought to bear and hence the output responses of the filter is the sum of the channel response to each individual impulse function. We may therefore conclude that the overall effect of the filter upon the train of impulse functions shown in Figure 5.2(a) is the sum of a series of sinc functions each centred upon the time interval of its respective impulse function. The individual sinc function for each impulse function is shown in Figure 5.2(b). This figure does not include the final output voltage, the sum of the series of time shifted sinc functions. However, it is clear from the idealised channel output response

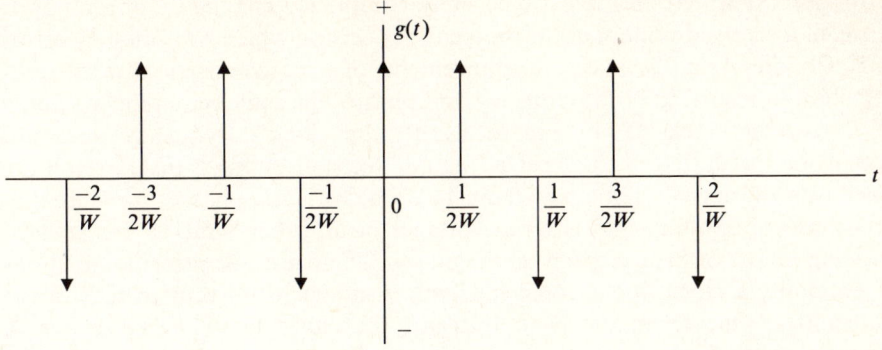

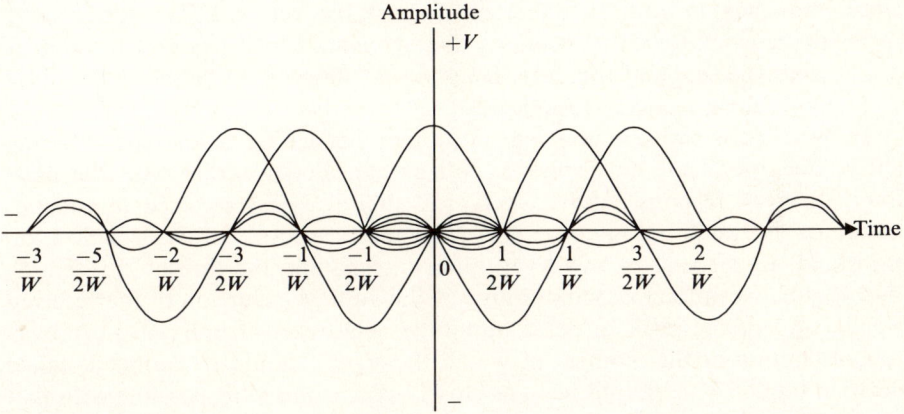

Figure 5.2   Transmission of a digital signal through ideal low-pass filter: (a) input signal; (b) output response of filter

shown that at the sampling intervals ($1/2W$, or multiples thereof) the amplitude is a peak and due entirely to the corresponding impulse pulse to the line. Secondly, the amplitude of all other sinc pulses at each sampling interval other than its own is zero. The receiver clock must be perfectly synchronised such that a decision regarding the signal's amplitude is made at the correct interval in time. Under this condition the only voltage present is that purely due to the desired received symbol in which case no other symbols interfere. This condition means that there is no **inter-symbol interference** (ISI).

The analysis has revealed that, even with a band-limited channel, each symbol may be correctly received. In practice digital signals are broader than impulse functions, leading to a smearing of the received signal. Secondly the channel response is imperfect, which introduces distortion effects, of which phase distortion is most detrimental. Thirdly, clock timing, and hence decision timing of the received signal, is subject to variation or **jitter**. All three effects introduce ISI which, in the extreme, may completely change the meaning, or sense, of a received symbol at the moment of decision which will cause an error.

In Chapter 4 we discussed the requirements of a receiver carrier for coherent detection in regard to both frequency and phase. The same requirements apply to a receiver clock. If, in Figure 5.2(b), the clock's frequency does not synchronise with that of the received signal, decision timing is lost. Even if the clock is of the correct frequency, it must also be correctly phased to ensure that decisions are optimal, that is when voltages of all other symbols are at zero-crossing points of their respective sinc pulses. There are two principal methods of obtaining a clock at the receiver which is in synchronism with that of the transmitter. One technique is to transmit the clock to the receiver over a separate channel. More usually, by suitable coding of the digital signal prior to transmission, it is arranged that it contains a strong spectral content at the clock frequency, or a multiple thereof. A clock recovery circuit at the receiver recovers the clock from the incoming digital signal. Clock recovery techniques are beyond the scope of this text. Duck et al.[3] describe a number of popular coding and clock recovery techniques.

In practice a transmission line, although possessing a low-pass filtering effect, does not have an abrupt cut-off as considered earlier with the ideal low-pass filter response above. A typical digital signal received in practice might be as shown in Figure 5.3(a). The receiver must establish a decision threshold. In the case of two-level (binary) signalling the threshold is simply mid-way between the maximum range of the voltage excursion of the received signal, or half the peak to peak amplitude. The receiver must also establish reliable timing of the moment at which the signal should be sampled, called **decision timing**. This should be in as closely synchronised as possible with that of the transmitter's data clock.

A useful test performed in practice to monitor the effect of a transmission link, and associated noise and interference is that of an **eye diagram**. Consider again the received digital signal shown in Figure 5.3(a). If an oscilloscope is synchronised with the symbol rate such that each trace displays precisely one symbol, successive traces become superimposed as shown in Figure 5.3(b). For ease of illustration the symbols shown in Figure 5.3(a) have been numbered. There corresponding appearances within the eye diagram of Figure 5.3(b) are indicated. The exact number of traces visible will depend upon the persistence of the cathode ray tube of the cathode ray oscilloscope.

Figure 5.3(b) appears similar in shape to an eye, hence the name. The receiver samples the received signal periodically at the centre of each symbol, and hence eye. This is known as the decision moment. For a binary signal the decision threshold is set mid-way between peak positive and negative excursions of the received signal's voltage. A voltage which, at the decision moment, is above that

---

[3] Duck, M., Bishop, P. and Read, R., *Data Communications for Engineers*, Addison-Wesley, 1996, pp. 57–72. ISBN 0-201-42788-5.

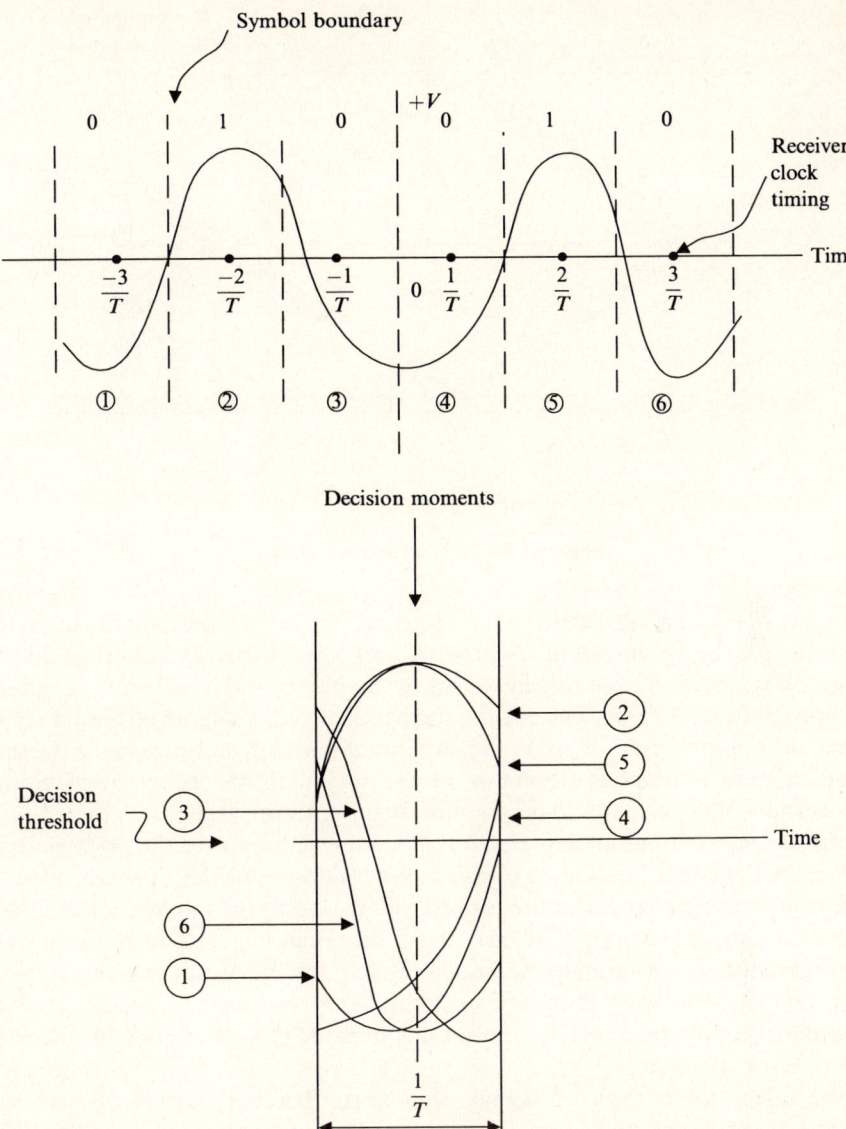

Figure 5.3 Eye diagram: (a) typical received digital signal; (b) eye diagram at receiver. Numbers in circles denote symbol numbers

of the threshold level is output as binary 1, and vice versa. It may be clearly seen from the eye diagram that the signal amplitudes at the moment of decision vary. This is an indication that ISI is present and which is caused by distortion, noise, interference and band limiting of the channel.

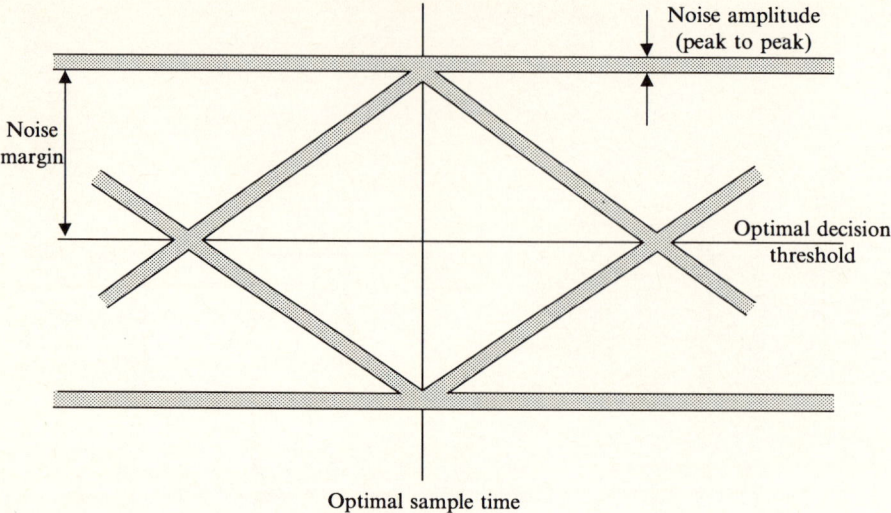

Figure 5.4 Generalised eye diagram

Figure 5.4 shows a generalised eye diagram. The fuzzy bands are the lines due to signal, as shown earlier in Figure 5.3, and accompanying variation due to noise. Noise increases the overall width of the bands and its effect is to reduce the open area of the eye. The eye diagram indicates the degree of immunity to noise, or noise margin. This is the amount by which noise may be further increased before complete closure of the eye occurs. If this occurs, no meaningful decision may be made and reception becomes impossible. In practice, the decision level itself has an associated tolerance which reduces the noise margin further. An eye that has a wide opening horizontally, or sides of small slope, is tolerant to timing error variation, or **jitter**, in regard to the moment of decision. Under this condition any jitter only has a marginal effect upon noise margin and error rate is substantially unaltered. Where the eye is narrow and slope is steep, tolerance to jitter diminishes rapidly as the moment of decision moves from the optimum position. Thus eye slope indicates tolerance, or otherwise, to timing error or jitter.

The effect upon received signal of a more gradual roll-off of practical filtering is apparent in Figure 5.3. An **equaliser** may be used to reduce the amplitude of oscillatory tails and to preserve the zero-crossing characteristic of sinc pulses. The combined response of channel and equaliser can be made equivalent to multiplication in time an of ideal low-pass filter response of bandwidth equivalent to the Nyquist minimum with a suitable frequency response to satisfy the above requirements. We shall show shortly that an ideal low-pass filter response convolved by a frequency response which exhibits even symmetry produces a time response with zero-crossing characteristics similar to that of a sinc function. A popular response which is sought in practice is that of

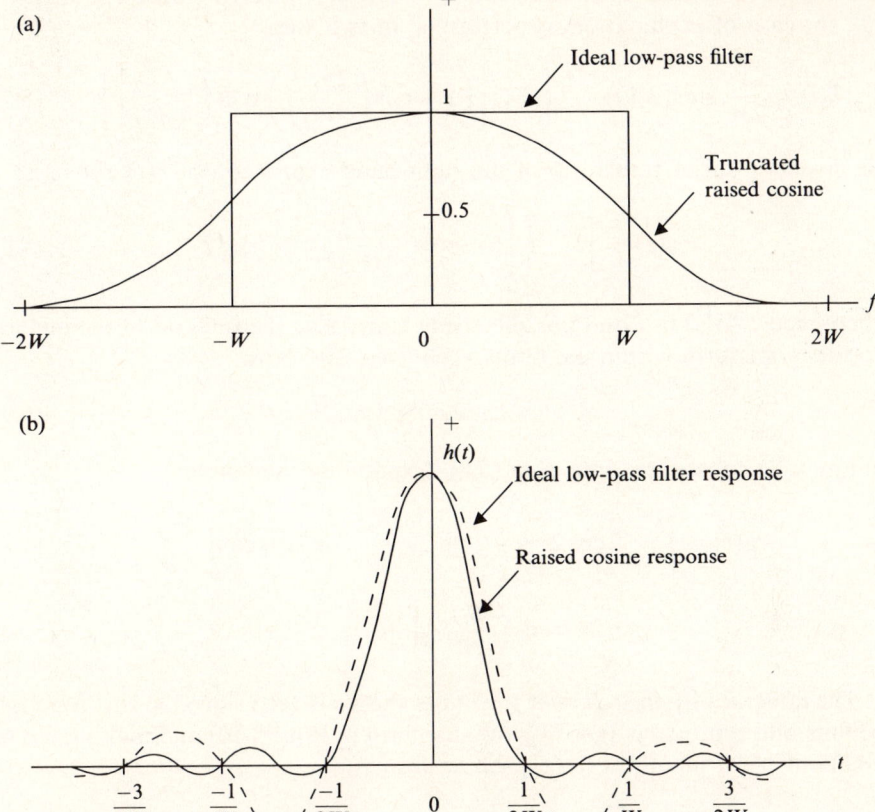

Figure 5.5 Raised cosine channel response: (a) frequency response; (b) impulse response

a truncated raised cosine[4]. Figure 5.5(a) illustrates the ideal and truncated raised cosine frequency responses desired.

Let the desired channel and equaliser response be $H(f)$ which is equivalent to the convolution of a rect function, representing the ideal response, with a raised cosine response with twice the bandwidth. This may be achieved by multiplication of the raised cosine function by a rect function and the overall expression for $H(f)$ is shown:

$$H(f) = \text{rect}\left(\frac{f}{2W}\right) * \frac{1}{2}\left\{\left[1 + \cos\left(\frac{2\pi f}{4W}\right)\right]\text{rect}\left(\frac{f}{4W}\right)\right\} \qquad (5.6)$$

[4] Stremler, F.G., *Introduction to Communications Systems*, 3rd edn, Addison-Wesley, 1990, pp. 386–90. ISBN 0-201-51651-0.

The impulse response is found by taking the inverse Fourier transform of $H(f)$ but, for ease of explanation, is performed in two steps:

$$h(t) = \frac{1}{2W}\operatorname{sinc}(2Wt) \cdot \frac{1}{2}\left\{\mathcal{F}^{-1}\left[\left(1 + \cos\left(\frac{2\pi f}{4W}\right)\right)\operatorname{rect}\left(\frac{f}{4W}\right)\right]\right\} \qquad (5.7)$$

the inverse Fourier transform of the right-hand expression of $h(t)$ above is:

$$g(t) = \int_{-2W}^{2W}\left\{\left[1 + \cos\left(\frac{2\pi f}{4W}\right)\right]e^{j2\pi ft}\right\}df \qquad (5.8)$$

where $\operatorname{rect}(f/4W)$ in Eqn (5.6) effectively constrains the integral of the inverse Fourier transform within the limits $-2W$ to $+2W$. Now:

$$e^{j2\pi ft} = \cos 2\pi ft + j\sin 2\pi ft \qquad (5.9)$$

and may be substituted into Eqn (5.8). Expanding then gives:

$$g(t) = \int_{-2W}^{2W}\left[\cos 2\pi ft + j\sin 2\pi ft + \cos 2\pi ft\cos\left(\frac{2\pi f}{4W}\right)\right.$$
$$\left. + j\sin 2\pi ft\cos\left(\frac{2\pi f}{4W}\right)\right]df \qquad (5.10)$$

The integral of $j\sin 2\pi ft$ over the limits shown is zero since the sine function exhibits odd symmetry. Ignoring the sine term in Eqn (5.10) and making use of trigonometrical identities for a product results in:

$$g(t) = \int_{-2W}^{2W}\left\{\cos 2\pi ft + \frac{1}{2}\left[\cos\left(2\pi f\left(\frac{4Wt - 1}{4W}\right)\right)\right.\right.$$
$$\left. + j\sin\left(2\pi f\left(\frac{4Wt - 1}{W}\right)\right) + \cos\left(2\pi f\left(\frac{4Wt + 1}{4W}\right)\right)\right.$$
$$\left.\left. + j\sin\left(2\pi f\left(\frac{4Wt + 1}{W}\right)\right)\right]\right\}df \qquad (5.11)$$

Rearranging Eqn (5.11):

$$g(t) = \int_{-2W}^{2W}\left[\cos 2\pi ft + \frac{1}{2}\cos\left(\frac{4Wt - 1}{2W/\pi}\right)f + \frac{1}{2}\cos\left(\frac{4Wt + 1}{2W/\pi}\right)f\right]df \qquad (5.12)$$

$$= \frac{1}{2}\left[\frac{1}{\pi t}\sin 2\pi ft + \frac{2W/\pi}{4Wt - 1}\sin\left(\frac{4Wt - 1}{2W/\pi}\right)f\right.$$
$$\left. + \frac{2W/\pi}{4Wt + 1}\sin\left(\frac{4Wt + 1}{2W/\pi}\right)f\right]_{-2W}^{2W} \qquad (5.13)$$

and, remembering that $\sin(-x)$ equals $-\sin(x)$:

$$g(t) = \frac{1}{\pi}\left[\frac{1}{t}\sin 4W\pi t + \frac{2W}{4Wt-1}\sin(4Wt-1)\pi\right.$$

$$\left. + \frac{2W}{4Wt+1}\sin(4Wt+1)\pi\right] \tag{5.14}$$

$$= \frac{1}{\pi}\left\{\frac{1}{t}\sin 4W\pi t + \frac{2W}{4Wt-1}\right.$$

$$\times [\sin 4W\pi t\cos(-\pi) - \cos 4W\pi t\sin(-\pi)]$$

$$\left. + \frac{2W}{4Wt+1}(\sin 4W\pi t\cos\pi + \cos 4W\pi t\sin\pi)\right\} \tag{5.15}$$

Since $\cos\pm\pi$ equals -1 and $\sin\pm\pi$ equals 0, Eqn (5.15) may be rewritten:

$$g(t) = \frac{1}{\pi}\left(\frac{1}{t}\sin 4W\pi t - \frac{2W}{4Wt-1}\sin 4W\pi t - \frac{2W}{4Wt+1}\sin 4W\pi t\right) \tag{5.16}$$

$$= \frac{1}{\pi}\sin 4W\pi t\,\frac{1}{t(1-4^2W^2t)} \tag{5.17}$$

Therefore Eqn (5.17) may be substituted in Eqn (5.7) and hence the impulse response as seen at the output of the equaliser is:

$$h(t) = \operatorname{sinc} 2Wt\,\frac{1}{4W\pi t}\,\frac{\sin 4W\pi t}{(1-4^2W^2t)} \tag{5.18}$$

$$= \operatorname{sinc} 2Wt\,\frac{\sin 4W\pi t}{4W\pi t(1-4^2W^2t^2)} \tag{5.19}$$

$$= \operatorname{sinc} 2Wt\,\frac{\sin 4W\pi t}{(1-4^2W^2t^2)} \tag{5.20}$$

An impulse produces a time response at the equaliser output as shown in Figure 5.5(b) where we may see that, as with a sinc response of an ideal channel, there is zero ISI. Raised cosine pulse shaping is slightly less tolerant to timing jitter because of the narrower pulse but the tails decay far more rapidly offering superior ISI performance. In practice broader pulses that contain more energy are employed to provide greater immunity to noise. The equaliser response is made to produce the same overall response as that required for an impulse. Hence the time domain response to a finite width pulse remains the same.

5.1.2 Effect of noise at the receiver

We shall now consider the effect of Gaussian noise, which is the most common form of noise in communication systems, upon a digital signal introduced during transmission. In the case of binary signalling the situation at the receiver may be represented by Figure 5.6 where two symbols, A and B, are shown. The probability density function (PDF) for Gaussian noise is superimposed upon the nominal signal levels V_A and $-V_B$.

If, when symbol A is being transmitted, the associated noise voltage amplitude is less than $-V_A$ at the moment of decision, the symbol will be interpreted as that of symbol B and an error produced. Let symbol A being transmitted be defined as event A and the noise being such that the received voltage is less than 0 V, and hence an error occurring, be event B. We may then say that the probability of an error occurring when symbol A is transmitted is the *joint probability* of these two events written as $P(AB)$. We also know that there is a *conditional probability* in that an error may, in the above case, only occur providing symbol A is transmitted. This conditional probability is written as $P(B|A)$ and means the probability of the noise being less than $-V_A$, given that symbol A is transmitted. $P(AB)$ and $P(B|A)$ are related through the

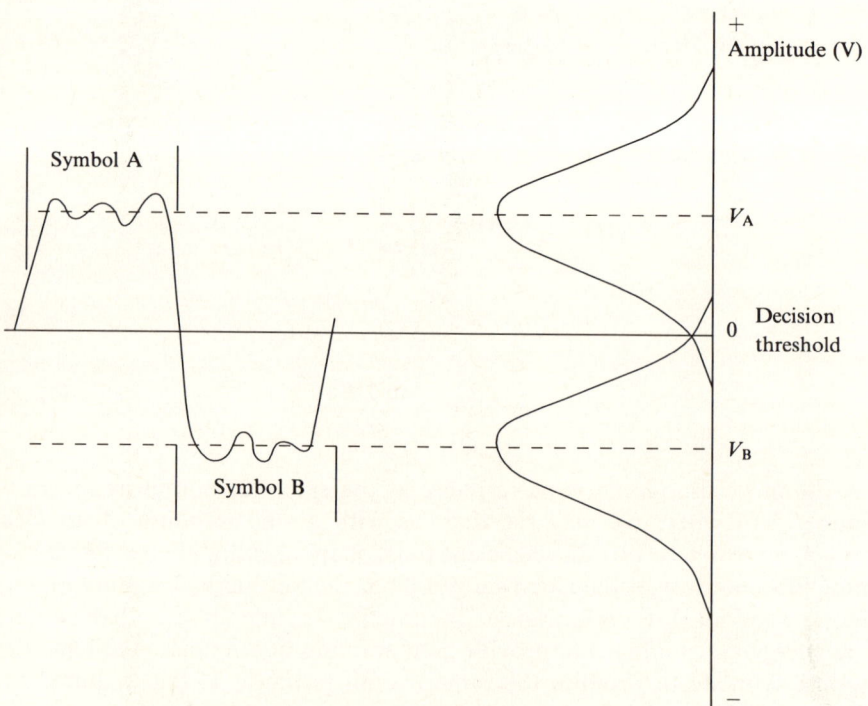

Figure 5.6   Effect of noise upon a binary signal

well-known expression:

$$P(B \mid A) = \frac{P(AB)}{P(A)} \qquad (5.21)$$

Hence $P(A \mid B)$, the probability of an error if symbol A is transmitted, is given by:

$$P(AB) = P(B \mid A)P(A) \qquad (5.22)$$

In a similar manner we may deduce that the probability of an error when symbol B is transmitted is given by:

$$P(BA) = P(A \mid B)P(B) \qquad (5.23)$$

We shall assume that symbols A and B are equiprobable, hence $P(A)$ and $P(B)$ each have a probability of 0.5 and that the threshold is set midway between voltage levels A and B. (If symbols A and B are not equiprobable a scrambler may be used prior to transmission to randomise them and ensure they are equiprobable.) Therefore the overall probability of an error P_e is given by:

$$P_e = P(A \mid B)P(B) + P(B \mid A)P(A) \qquad (5.24)$$

We may reasonably assume that $P(A)$ and $P(B)$ are equiprobable, each of probability 0.5. The symmetry of the Gaussian PDF means that $P(A \mid B)$ and $P(B \mid A)$ are also equal. Hence we may readily deduce:

$$P_e = P(A \mid B) \qquad (5.25)$$

$$= P(B \mid A) \qquad (5.26)$$

Let the potential difference between V_A and $-V_B$ equal 2 V. Hence P_e may be expressed:

$$P_e = \int_V^\infty P_x(x)\,\mathrm{d}x \qquad (5.27)$$

That is, error probability is simply the probability that the noise amplitude exceeds $+V$ and is known as the **tail probability** of a PDF. Now $P_x(x)$, the noise probability distribution function, is in this case the Gaussian PDF, shown in Figure 5.7.

In the case of a Gaussian distribution, $P_x(x)$ in Eqn (5.27) is widely published and the equation may be written:

$$P_e = \frac{1}{\sigma\sqrt{(2\pi)}} \int_V^\infty \exp\left(\frac{-x^2}{2\sigma^2}\right) \mathrm{d}x \qquad (5.28)$$

where, in this case, σ is equivalent to the rms value of the noise voltage.

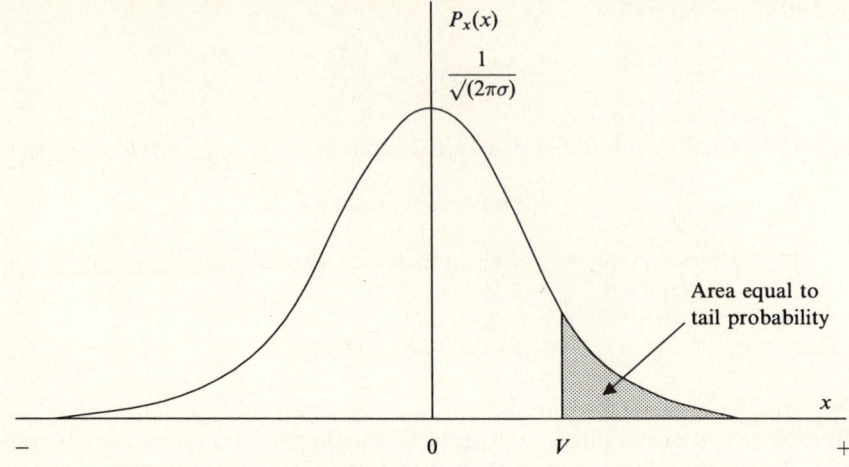

Figure 5.7   Gaussian PDF

The probability of an error is also known as **bit error rate** (BER) and, for example, a BER of 10^{-6} simply means that on average one bit will occur in every million transmitted. Where noise and symbol amplitudes are known, error probability may not be found directly from Eqn (5.28) because the integral is not in closed form. A number of methods may be used to perform numerical calculations relating to BER, noise and symbol amplitude but we shall confine ourselves here to use of the **error function**[5], or erf(x):

$$\mathrm{erf}(x) = \frac{2}{\sqrt{\pi}} \int_0^x e^{-u^2}\, du \qquad (5.29)$$

where:

$$x = \frac{V}{\sigma\sqrt{2}} \qquad (5.30)$$

The above expression for error function is equivalent to the probability that x lies in the range $0 \leqslant x \leqslant \pm V$. Error probability equals the area under the curve for $x > V$. Now the area under the curve of the PDF for all x is of course 1. Hence the area under the curve to the right of V is given by:

$$P_e = 0.5[1 - \mathrm{erf}(x)] \qquad (5.31)$$

---

[5] Roden, M.S., *Analog and Digital Communication Systems*, 4th edn, Prentice Hall, 1996, pp. 113–14. ISBN 0-13-399965-3.

In the case of two symbols separated by $2V$ mean signal power to noise power equals:

$$\frac{\text{Mean signal power}}{\text{Noise power}} = 20 \log_{10} \frac{V}{\sigma} \text{ dB} \qquad (5.32)$$

EXAMPLE 5.1
A telegraph channel employs dc signalling using two symbols whose voltages are $+80$ V and -80 V. If the S/N ratio in the presence of Gaussian noise is 8 dB, estimate the resulting bit error rate.

SOLUTION
From Eqn (5.32) we may write:

$$\frac{V}{\sigma} = a \log_{10} \frac{\text{mean signal power}}{20 \times \text{noise power}} \qquad (5.33)$$

Hence:

$$\frac{V}{\sigma} = a \log_{10} \frac{8}{20} \qquad (5.34)$$

$$= 2.512 \qquad (5.35)$$

Substituting the value for V/σ into Eqn (5.30) we find:

$$x = \frac{2.512}{\sqrt{2}} \qquad (5.36)$$

$$= 1.776 \qquad (5.37)$$

We may now determine $\text{erf}(x)$ from the table in Appendix E. An estimated value for $\text{erf}(x)$ is 0.988. Hence, from Eqn (5.31) BER is found to be:

$$\text{BER} = 0.5(1 - 0.988) \qquad (5.38)$$

$$= 6 \times 10^{-3} \qquad (5.39)$$

So, in this example, BER is between 1 in 100 and 1 in 1000. For such a poor S/N ratio, such a low value of BER is not unsurprising.

Self-assessment 5.1 A baseband digital system is to employ binary signalling over a channel which produces a noise (assumed Gaussian) at the input terminals to the receiver of 100 mV. If BER is to be better than 10^{-5} determine the minimum voltage separation between the voltage level of each symbol.

Instead of binary signalling, multi-level signalling may be employed where one of three, or more, symbols may be transmitted. For instance four symbols could represent pairs of bits, known as **dibits**, 00, 01, 10 and 11. Such signalling enables the line signalling rate to be reduced compared with the **transmission rate** at which bits are sent. In the case of four symbols four different voltages are necessary, one to represent each symbol. Transition between symbols only occurs after every two bits, hence the signalling rate is half that of the transmission rate. This simple illustration serves to indicate the difference between transmission rate, measured in bit per second and signalling rate measured in baud. Note that in the case of binary signalling, transmission rate and signalling rate are equal.

If more symbols are employed within the same overall range of voltage their spacing becomes closer than in the case of binary signalling. Therefore, for a given noise level, error probability increases. Analysis reveals that to maintain the same error probability as for the binary case, S/N must be raised by:

$$10 \log_{10} \frac{m^2 - 1}{3} \ \mathrm{dB}$$

where m represents the number of symbols employed.

If noise remains the same, the additional S/N required ensures adequate spacing between symbol levels for a specified BER. This means that the overall range of voltage employed is increased compared with binary signalling for the same error rate.

In practice systems often consist of a series of links in tandem. The overall BER of such systems equals the sum of the BER rate of each individual link, providing that the noise in each link is uncorrelated with that of any other link, which is generally the case.

5.1.3 Channel capacity

The maximum transmission rate at which information may be conveyed over a channel with an arbitrarily small error rate is called the **channel capacity**. Communication channels are less than ideal in that bandwidth is limited and noise is present. **Shannon's channel capacity** theorem relates these factors to capacity:

$$C = W \log_2 \left(1 + \frac{S}{N} \right) \text{ bit per second (bps)} \tag{5.40}$$

where: C is the channel capacity,
         W is the bandwidth,
         S/N is the signal to noise ratio,

and where noise is assumed to have a Gaussian distribution. Note that S/N is a ratio, not expressed in decibel.

A casual glance at Eqn (5.40) might suggest that channel capacity increases without bound with both S/N and bandwidth. This is somewhat misleading. In practice signal power is limited and Gaussian noise, as we saw in Chapter 3, equals the product ηW. For a given bandwidth channel capacity does increase with S/N. Now consider the effect of holding signal level constant and increasing bandwidth. This means that noise power also increases linearly with bandwidth, causing a logarithmic reduction in S/N. Therefore although channel capacity increases with bandwidth, its rate of increase declines as S/N becomes reduced, owing to increased noise power with bandwidth. In the limit, as bandwidth approaches infinity, channel capacity is bounded by the asymptotic value of $S/N \log_2 e$, or 1.44S/N. This is sometimes called the **Shannon boundary**. In practice effects such as crosstalk (see below) and distortion reduce channel capacity further.

5.2 Time division multiplexing

Figure 5.8 illustrates the principle of time division multiplexing (TDM). TDM is the process of switching a number of signal sources, or channels, in strict rotation on a one at a time basis to a single output. Three such channels are shown in the figure. A device that performs this function is known as a multiplexer. Providing each input channel is sampled sufficiently rapidly to satisfy the Nyquist criterion discussed in Chapter 2, each individual signal is perfectly recoverable at a receiver. In practice such an analogue-based TDM system as shown would suffer considerable distortion during transmission due to band-limiting effects, especially of signal edges between adjacent samples. The figure is only for purposes of illustration and would not be used in practice.

Modern TDM systems are digital and multiplex binary signals which may come from a computer. Other types of signal such as telephony or video, which are analogue in nature, may be digitised by means of an ADC prior to multiplexing. Multiplexers usually have four input channels for efficient use of two control lines that govern which particular input is to be switched to the output at any particular instant in time.

Multiplexers employ readily available digital logic-based circuits and are therefore cheap. The receiver must perform a complementary function to that of the multiplexer known as demultiplexing. The receiver clock controlling the demultiplexer operation must be very carefully synchronised with that of the incoming TDM signal to ensure that switching occurs exactly on the transition between received signal elements to avoid signals, or fragments of signals, passing from one channel into another. The effect of such distortion is generally known as **crosstalk**.

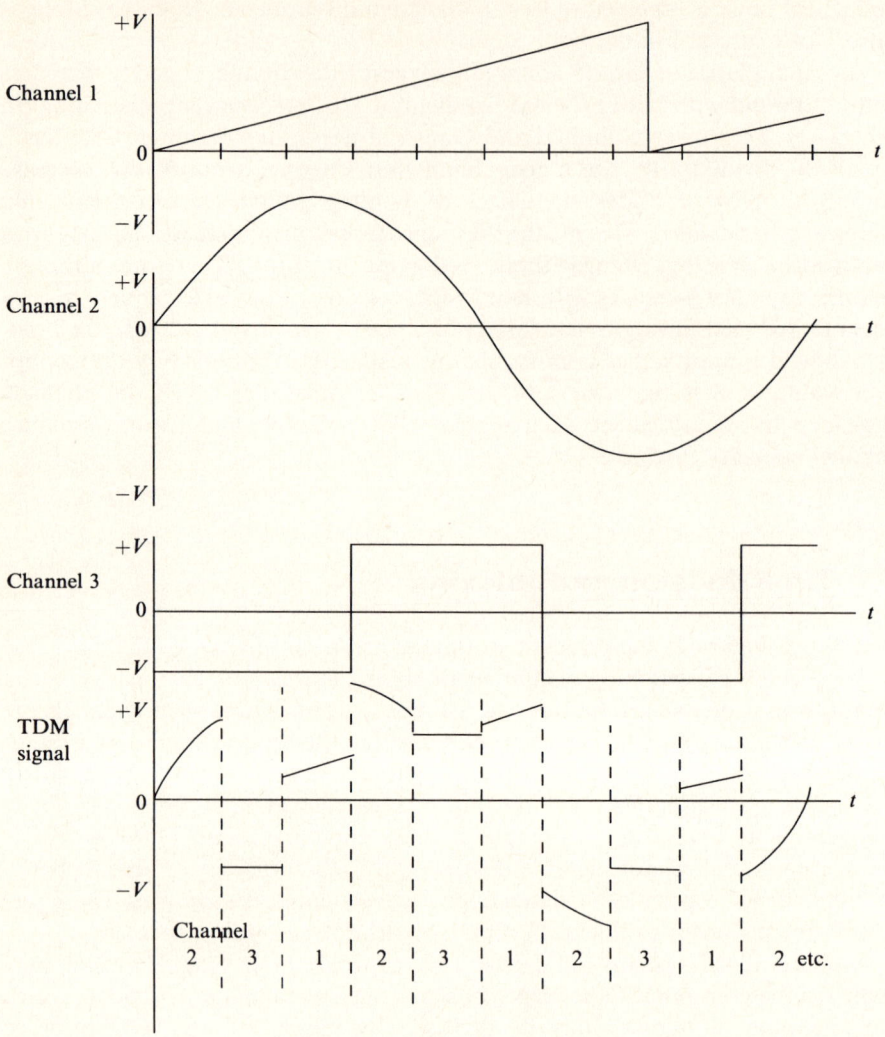

Figure 5.8   Time division multiplexing

Self-assessment 5.2 A four-way multiplexer has, over a period of time equal to four input bits, the following bit sequences on its input channels: Channel A 1001, Channel B 1100, Channel C 0101 and Channel D 0010. Sketch carefully to a common time axis, in a manner similar to Figure 5.8, the signals of each input channel and the associated multiplexed waveform.

Signals at the output of a multiplexer are of higher rate than that of each of its input channels. It is assumed that each input channel is of the same transmission rate. In the case of a four-way multiplexer during each input bit period, the multiplexer must output four bits, one bit each of the input channels. It is therefore evident that the transmission rate of a multiplexed signal is equal to the product of the number of input channels and their associated transmission rate. By means of a succession of multiplex stages, signals of very high transmission rate may be produced. In consequence, one of the advantages of TDM is the ability to make use of the high bandwidth of physical media now available for transmission, especially optical fibre. Commercial transmission in excess of 1 Terabit per second (Tbps) has recently become available. Such high transmission rates enable large numbers of separate signals to be carried over the same physical medium, rather than a plethora of separate physical paths. This in turn leads to savings in space, and cost, of transmission media.

5.3 Pulse code modulation

Pulse code modulation (PCM) in simplest form is the conversion of an analogue signal into digital form for onward transmission. A signal is sampled at regular intervals in time and converted to a digital signal by means of an ADC. Each sample is then transmitted serially, rather than in parallel form, over a single path as a series of bits. Each set of bits represents a single sample and may be regarded as a **pulse code**, hence the signal is known as pulse code modulated.

Figure 5.9 illustrates a 30-channel PCM system widely used in telephone systems throughout Europe. The transmission equipment that performs the necessary complex electronic signal processing tasks has traditionally been uneconomic to provide for a single circuit or channel (although this is now changing). It may be seen how TDM is used to share a single ADC, channel and DAC. Note that only one direction of transmission is shown. In practice,

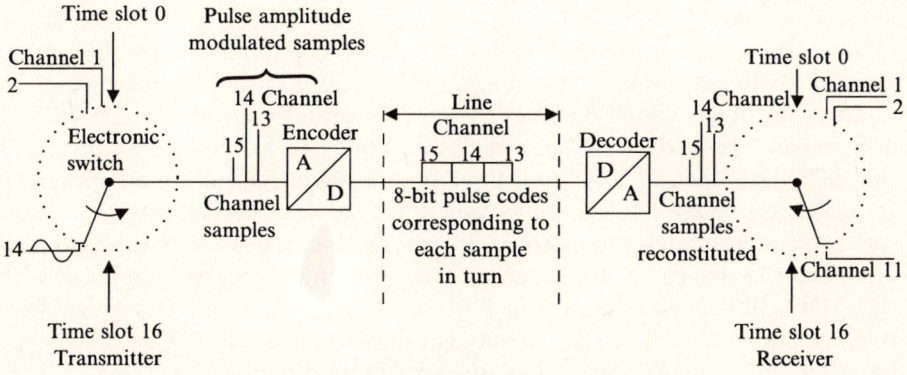

Figure 5.9 30-channel pulse code modulation system

to support two-way communication, an identical arrangement also exists in the reverse direction. Each channel is assumed to have a bandwidth of 300 Hz to 3.4 kHz (the bandwidth widely accepted for commercial quality speech). Each input channel is sampled in turn in analogue form at a frequency of 8 kHz to satisfy the Nyquist criterion. The resultant **Pulse amplitude modulated** (PAM) signal is applied to an encoder where each sample is converted into an 8-bit word, or pulse code, by means of an ADC. There are thus 256 different voltage, or **quantisation levels**, possible. In practice only 30 time slots are allocated for speech channels. The remaining two time slots, numbered 0 and 16, are used for frame alignment and common channel signalling. Frame alignment ensures that the multiplexer/demultiplexer pair runs in exact synchronism time slot for time slot. Signalling enables information such as dialling information and clear-down signals for all 30 channels to be sent over a common channel using a single time slot. (This is known as **common channel signalling** and is very efficient.)

The receiver performs the converse functions for reception, namely decoding and demultiplexing. In addition each output channel includes a low-pass filter (not shown) known as a **reconstruction filter**, with a cut-off frequency of 4 kHz, to convert the recovered PAM samples into analogue form.

PCM was developed for short-haul telephone links among exchanges. Twenty-four (for example in Japan and the USA) or 30 simultaneous telephone circuits may be accommodated over two standard telephone pairs. PCM enabled the rapid growth in telephone traffic during the 1960s to be met without the need to lay large numbers of additional cables. A 30-channel system offers a possible 15-fold saving in line plant, saving up to 26 pairs. The transmission speed of a single speech channel is effectively 64 kbps. Subsequently, 64 kbps has become a universal standard throughout the telecommunication industry and is now regularly used to carry other non-voice services, especially data.

Figure 5.10 illustrates the principle used in PCM for encoding and decoding. With reference to Figure 5.10(a), the ADC has $(M - 1)/2$ plus 1 (including $+0$) positive possible output states, or quantisation levels, and $(M - 1)/2 + 1$ negative levels, that is a total of $M + 1$ levels. Note that in this representation of the ADC characteristic, M is assumed to be an odd number. The question of why are there two possible zero output states arises. Consider the case of an input to the ADC which persistently remains within the range $\pm\Delta V$. If only one zero code existed, commonly an all-zero binary pattern would be output by the ADC. Absence of any transitions in the digital signal would present a problem at a receiver to determine where the boundaries are between binary sequences representing successive samples. That is clock recovery timing would be difficult to achieve at the receiver. The encoding algorithm employed with the ADC characteristic shown in Figure 5.10(a) is that over time the same number of positive and negative output values are produced. This is achieved by arranging that successive zeros are encoded with opposite sign, even if they are non-contiguous. This overcomes the above problem of a single zero code-

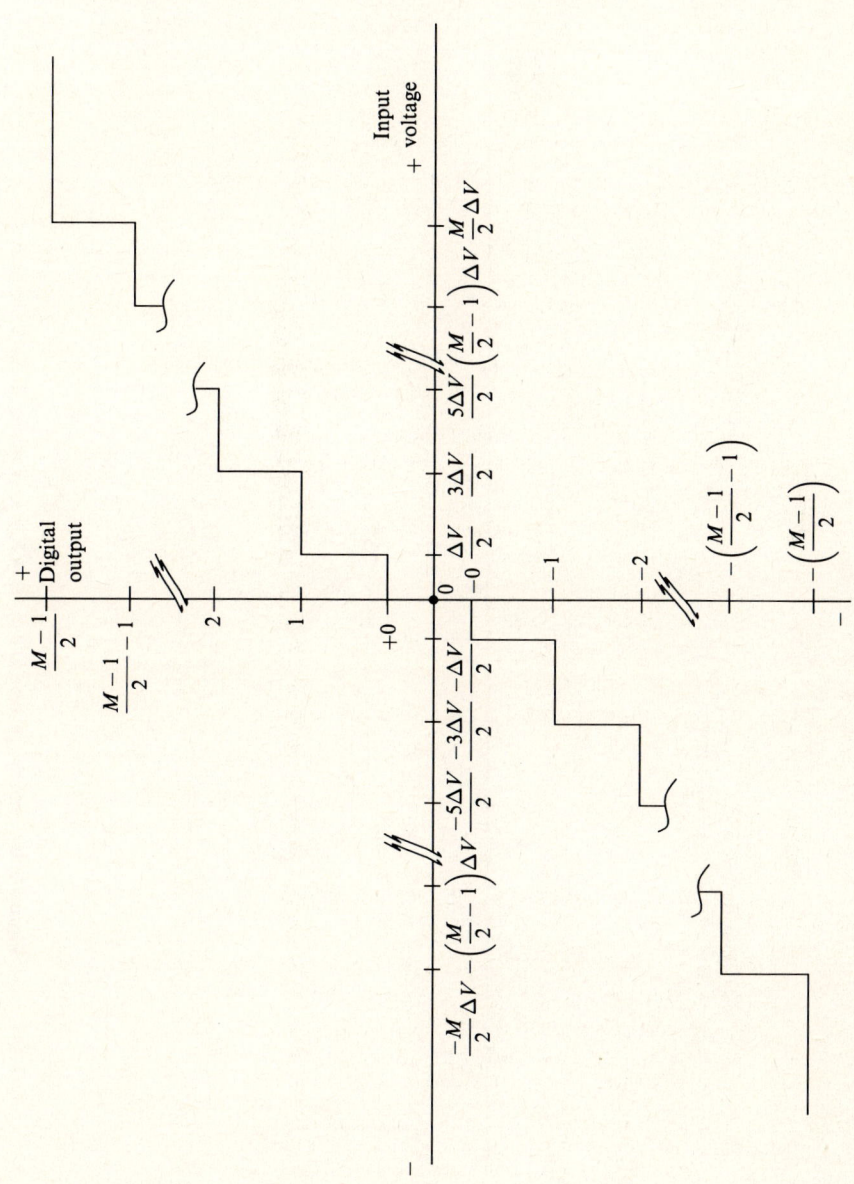

(a)

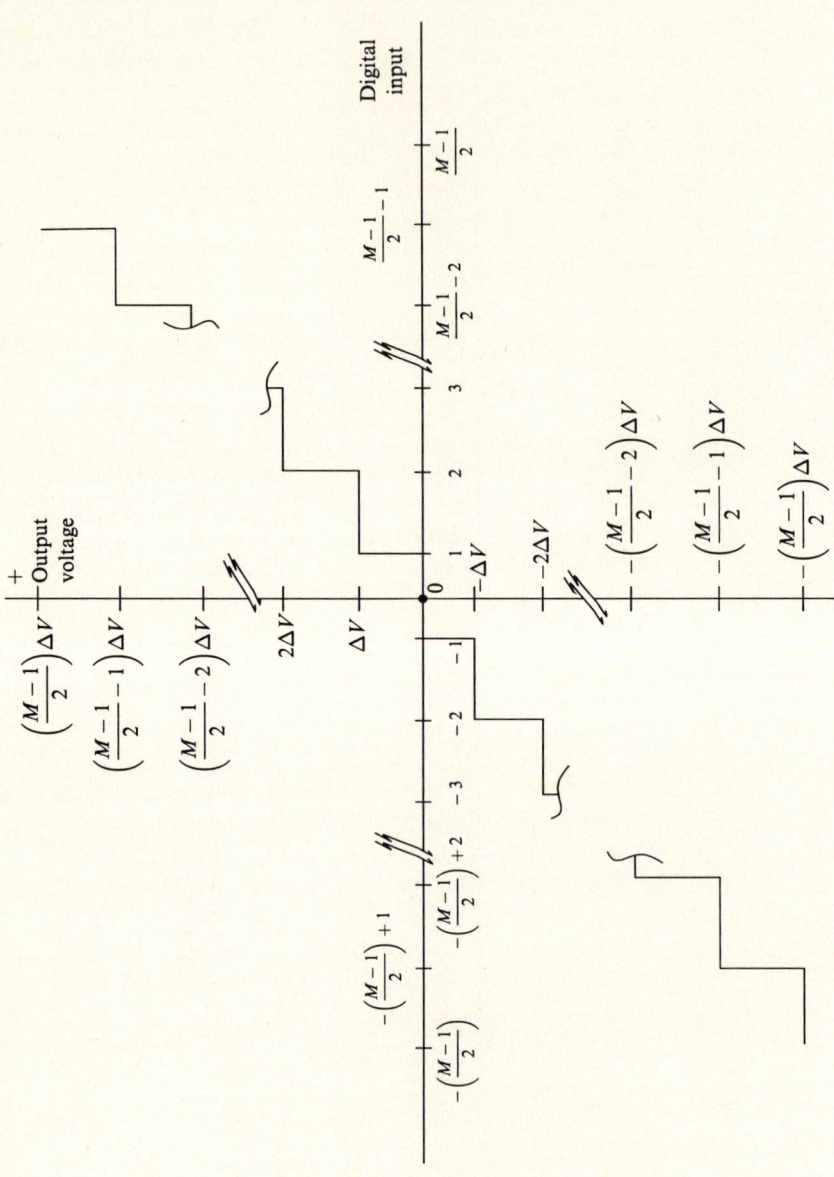

Figure 5.10 Encoding and decoding: (a) ADC characteristic; (b) DAC characteristic

word and has the advantage that, with suitable line coding prior to transmission over a channel, adequate clock timing information may be retained compared with sending, say, 0000 0000 repeatedly.

In examining Figure 5.10(a) further we see that as the input signal is progressively increased from zero amplitude to $+\Delta V/2$, the ADC output state is $+0$. Once an amplitude of $+\Delta V/2$ is reached 1 is output. Thereafter amplitude has to be increased by a further ΔV before 2 is produced, and so on. However, if amplitude ranges beyond $\pm M/2 \Delta V$ no further change in output value is possible and saturation occurs.

If N is the number of bits per sample, it follows from the above that:

$$M = 2^N - 1 \tag{5.41}$$

Typically 8-bit encoding is employed in which case there are 2^8, or 256, possible quantisation levels. It is usual to use the most significant bit of the output word to indicate voltage polarity. The remaining seven bits effectively represent the modulus of amplitude. For example, with an eight-bit ADC, 1XXX XXXX indicates a positive amplitude and 0XXX XXXX a negative amplitude.

Figure 5.10(b) shows the DAC characteristic located at the receiver. Zero signals, $+0$ or -0, are converted to $0V$. All other values, ± 1 through $\pm(M - 1)/2$ are converted to a corresponding voltage in the range $\pm\Delta V$ to $\pm(M - 1)/2$ times ΔV, respectively. What is particularly significant is the overall response of a PCM channel, as shown in Figure 5.11. Here we may see that any positive input voltage at the transmitter with amplitude between $0V$ and $\Delta V/2$ appears at the output of the system as 0. Voltages between $\Delta V/2$ and $\Delta 3V/2$ (a range of ΔV) appear at the output as ΔV, and so on. From this overall, or transfer, characteristic it is readily seen that there are discrete ranges over which the input may vary, equal to ΔV, but where no change in output voltage occurs. This means that in general there is a discrepancy between input voltage and output voltage, albeit small. Figure 5.12 illustrates this discrepancy which is known as **quantisation error**. It may be seen that the ADC and DAC characteristics have been carefully tailored such that the maximum value of quantisation error is $\pm\Delta V/2$, that is only half the quantisation step size of the ADC.

Quantisation error has profound implications in digital transmission. Consider a digital system based upon PCM. Transmission errors corrupt the value of a digital sample and may have a severe effect on a single sample. Providing such errors are relatively infrequent, that is far less than once per cycle of highest signal frequency component, they may be ignored in most cases. (Where this is not the case, particularly in transmission of data, error control techniques may be employed.)

It is possible to cascade a number of PCM systems where signals are converted to analogue between systems. However, the overall quantisation error remains within the limits $\pm\Delta V/2$. In digital systems noise is equivalent to quantisation error. Therefore quantisation noise is independent of distance.

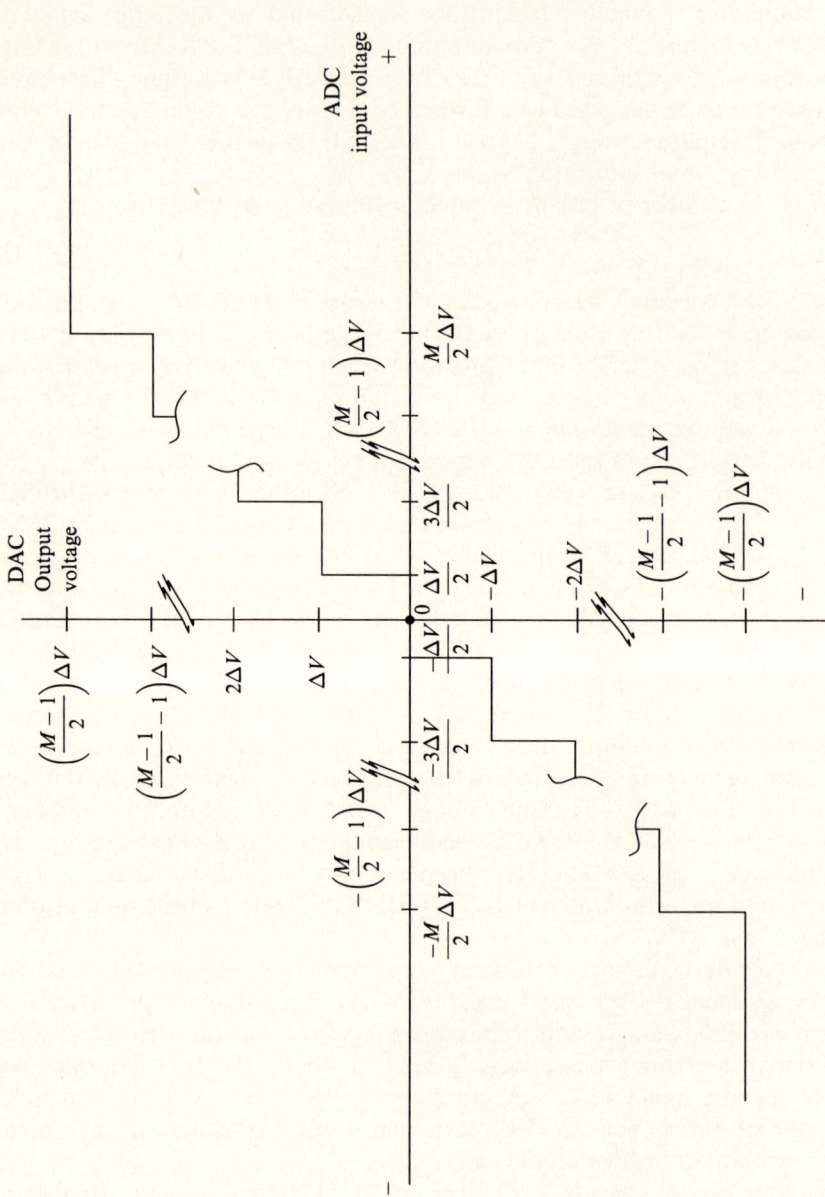

Figure 5.11 Overall ADC/DAC characteristic

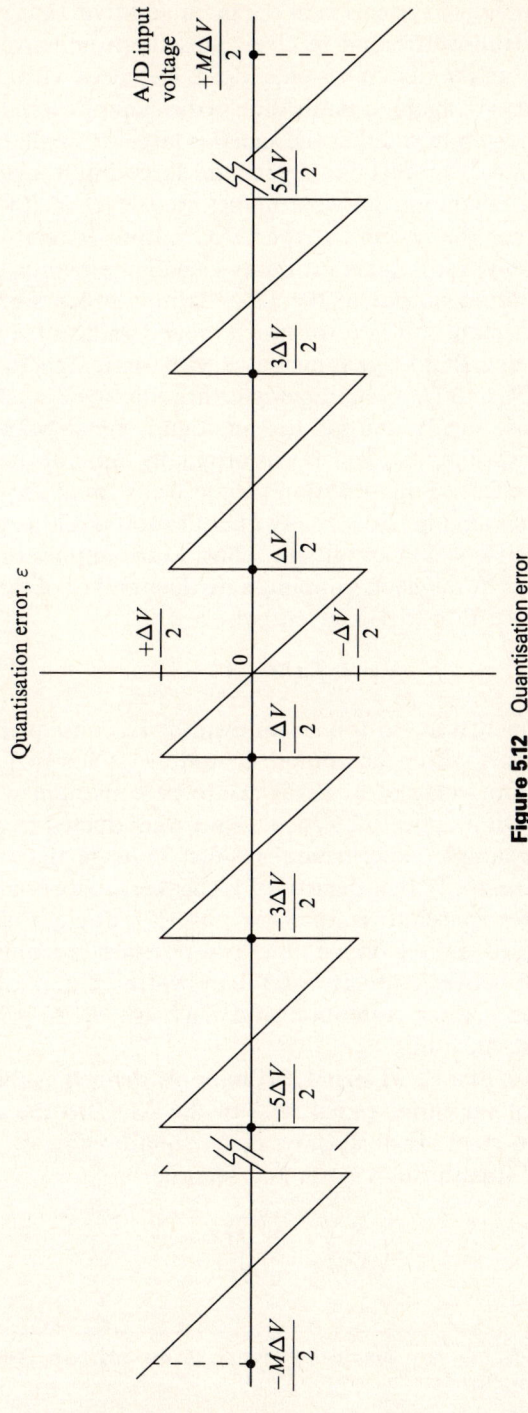

Figure 5.12   Quantisation error

However, noise in analogue systems is in the main additive; that is, as a signal propagates between transmitter and receiver, noise is progressively added, or increases. Therefore, unlike digital systems, noise increases with distance.

Looking at Figure 5.12 again, quantisation error cannot exceed $\pm\Delta V/2$ but its significance depends upon whether the signal is large or small. Consider, for instance, a speech signal. Loud talkers generate large input signals, perhaps peaking close to the maximum possible undistorted level of $[(M-1)]/2\Delta V$. Any quantisation error that occurs at the DAC output is relatively small in comparison. Conversely, a soft talker produces a small input voltage at the ADC input. The corresponding output of the DAC is now much smaller and the relative amplitude of quantisation error much more significant. Intuitively the signal to noise ratio in a PCM signal increases with input signal amplitude, at least where linear ADC and DAC characteristics are employed, as shown earlier.

A receiver should ideally output an analogue signal which faithfully matches, as near as possible, that which was originally input at the transmitter. This means that the effect of quantisation should be as small as possible. This may be achieved by arranging for as many quantisation levels as possible over the full range of possible signal amplitude. This in turn implies that M should be as large as possible. For a given sampling rate the number of bits per sample is, by rearrangement of Eqn (5.41), given by:

$$N = \log_2(M - 1) \tag{5.42}$$

Clearly, as M is increased, so too is bandwidth. In consequence, in order to constrain bandwidth within acceptable bounds, systems in practice must determine the minimum value of M necessary to be consistent with adequate quality transmission. In practice subjective tests are conducted to determine N, and hence M. For example experimental results[6] indicate that for adequate intelligibility (not necessarily the identity of a speaker), between 128 and 256 levels are necessary for speech. It is for these reasons that eight bits have been chosen for PCM used in telephone line transmission systems offering a reasonable trade-off between quality and bandwidth. Good quality music transmission requires greater resolution and may use up to 16 bits, which corresponds to 16-bit encoding.

Signal to noise ratio of a PCM signal is commonly defined as the ratio of the mean square value of the input signal $m(t)$ to the ADC to the mean square value of quantisation error. In this sense, mean square value of quantisation error may be termed quantisation noise N_q. That is:

$$\left(\frac{S}{N_q}\right) = \frac{\overline{m^2(t)}}{\overline{\varepsilon^2(t)}} \tag{5.43}$$

---

[6] Stremler, F.G., *Introduction to Communications Systems*, 3rd edn, Addison-Wesley, 1990, p. 544. ISBN 0-201-51651-9.

Before attempting to quantify signal to quantisation noise we need to derive an expression for $\overline{\varepsilon^2(t)}$. If, as is usual in practice, the peak value of ε is small in comparison with the peak permissible signal voltage, we may make the following assumption: all values of quantisation error over the full range of quantisation are equally likely. That is the PDF of ε, which we may call $P(\varepsilon)$, has a uniform distribution over a representative range $-\Delta V/2$ to $\Delta V/2$ as shown in Figure 5.13.

The mean square error may be found using the standard mathematical technique:

$$\overline{\varepsilon^2} = \frac{1}{\Delta V}\int_{-\Delta V/2}^{\Delta V/2} x^2\,\mathrm{d}x \tag{5.44}$$

$$= \frac{1}{\Delta V}\left(\frac{1}{3}x^3\right)_{-\Delta V/2}^{\Delta V/2} \tag{5.45}$$

$$= \frac{1}{3\Delta V}\left[\left(\frac{\Delta V}{2}\right)^3 - \left(\frac{-\Delta V}{2}\right)^3\right] \tag{5.46}$$

$$\therefore \quad \overline{\varepsilon^2} = \frac{\Delta V^2}{12} \tag{5.47}$$

In order to quantify signal to quantisation noise, the mean squared value of the signal voltage is required. In estimating the mean square noise voltage above, a uniform distribution for $P(\varepsilon)$ was quite reasonably assumed. However, the probability distribution of the signal voltage is far less certain. In practice signals are not uniformly distributed. The mean square signal voltage therefore depends upon the type of signal being carried and its associated

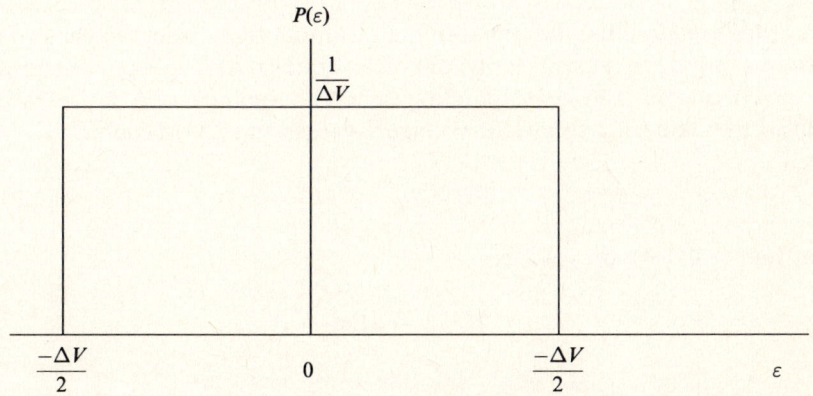

Figure 5.13 Error probability distribution

voltage distribution. Irrespective of knowledge of the precise voltage distribution for a given signal we may at least establish the signal to quantisation noise when the signal is at peak amplitude. The amplitude of the signal at the input to the ADC when it is at peak level may be expressed as:

$$\hat{V} = \frac{M}{2}\Delta V \tag{5.48}$$

From Eqn (5.47) we deduce that the rms value of noise voltage is:

$$\sqrt{\overline{\varepsilon^2}} = \sqrt{\frac{\Delta V^2}{12}} \tag{5.49}$$

$$= \frac{\Delta V}{\sqrt{12}} \tag{5.50}$$

Therefore, from Eqns (5.48) and (5.50), the ratio of peak signal voltage to rms noise voltage is given by:

$$\frac{\hat{V}}{\sqrt{\varepsilon^2}} = \frac{\dfrac{M}{2}\Delta V}{\dfrac{\Delta V}{\sqrt{12}}} \tag{5.51}$$

which, by rearrangement:

$$= \frac{M}{2}\Delta V \frac{\sqrt{4}\sqrt{3}}{\Delta V} \tag{5.52}$$

$$\therefore \quad \frac{\hat{V}}{\sqrt{\varepsilon^2}} = \sqrt{3}M \tag{5.53}$$

In order to determine the signal to quantisation noise ratio, we need to find the ratio of signal to noise in terms of powers, rather than voltage, as shown in Eqn (5.53) above. The corresponding signal to quantisation noise ratio is obtained by taking the square of voltage, hence Eqn (5.53) becomes:

$$(S/N_q)_{pk} = 3M^2 \tag{5.54}$$

From Eqn (5.42) we may write:

$$M - 1 = a\log_2 N \tag{5.55}$$

or:

$$M = 2^N - 1 \tag{5.56}$$

We may substitute Eqn (5.56) into (5.53):

$$\therefore \quad (S/N_q)_{pk} = 3(2^N - 1)^2 \tag{5.57}$$

or, in decibel form:

$$(S/N)_{pk} = 10\log_{10}[3(2^N - 1)^2] \, dB \tag{5.58}$$

$$\doteq 10\log_{10} 3 + 20\log_{10} 2^N \tag{5.59}$$

$$\doteq 4.8 + 6N \, dB \tag{5.60}$$

EXAMPLE 5.2

A PCM system encodes samples using an eight-bit ADC. In the case of peak signal amplitude, estimate the signal to quantisation noise ratio in decibel.

SOLUTION

$$(S/N_q)_{pk} \doteq 4.8 + 6N \, dB \tag{5.61}$$

In this example N is equal to 8. Therefore:

$$(S/N_q)_{pk} \doteq 4.8 + 6 \times 8 \, dB \tag{5.62}$$

$$= 52.8 \, dB \tag{5.63}$$

Clearly this is a very high value but is only valid for the peak amplitude signal voltage. Practical signals will have a correspondingly smaller value of signal to quantisation noise.

It is clear that Eqn (5.54), although a useful indication of the maximum possible signal to quantisation noise ratio, is not terribly meaningful for real signals. However, it forms a useful indicator of the signal to noise ratio for PCM systems.

The following self-assessment is included as an exercise to determine the signal to quantisation noise ratio on the assumption that signal voltage is uniformly distributed over the full range of permissible input voltage to the ADC.

Self-assessment 5.3 If a signal ranges between $-M\Delta V/2$ and $M\Delta V/2$, and all voltages within this range are equiprobable:

(a) show that the signal to quantisation noise ratio equals M^2
   (hint: produce, and make use of, a figure similar to that of Figure 5.13, but in terms of signal voltage, rather than error voltage);
(b) estimate signal to quantisation noise ratio in decibel if 8 bits/sample are employed.

Now let us consider the bandwidth requirement of a PCM signal. We saw earlier in this chapter that the maximum signalling rate possible (for binary signals) is 2 bit/Hz of bandwidth. Example 5.3 illustrates how bandwidth is related to the number of bits per sample and sampling rate.

EXAMPLE 5.3

Determine the minimum bandwidth required for the 30-channel PCM system discussed earlier. Compare this bandwidth with an equivalent FDM system.

SOLUTION

$$\text{Signalling rate} = \text{no. of time slots} \times \text{number of bits}$$
$$\text{per sample} \times \text{sampling rate} \tag{5.64}$$

A 30-channel PCM system uses 32 time slots in total. Therefore:

$$\text{Signalling rate} = 32 \times 8 \, \text{bits} \times 8000 \, \text{kHz} \tag{5.65}$$
$$= 2.048 \, \text{Mbps} \tag{5.66}$$

Therefore, the minimum bandwidth for PCM operation is 1.024 MHz.

An equivalent 30-channel FDM system, assuming each channel has a bandwidth of 4 kHz, is found thus:

$$\text{Bandwidth} = \text{no. of channels} \times \text{bandwidth of one channel} \tag{5.67}$$
$$= 30 \times 4 \, \text{kHz} \tag{5.68}$$
$$= 120 \, \text{kHz} \tag{5.69}$$

In comparison with an FDM system, a digital transmission system requires a bandwidth one order greater than that of an equivalent analogue system. PCM, as with FM, is an example of a bandwidth expansion system.

In the same way that we examined the effect of a receiver in demodulation of AM and FM in Chapter 4, it is possible to make similar comparisons for a PCM receiver. Consider Figure 5.14. We shall assume that the detector is ideal in that the information content at the input is the same as that at the output. Furthermore, we shall also assume that the input signal has a transmission bandwidth, B_T Hz, equal to that of the channel. As usual, the received message signal $m(t)$ at the output is of bandwidth W Hz. By means of Eqn (5.40) for channel capacity we may, from an information viewpoint, assume that no noise is lost in reception. Hence we may equate input and output conditions as follows:

$$B_T \log_2[1 + (S/N)_R] = W \log_2[1 + (S/N)_{out}] \tag{5.70}$$

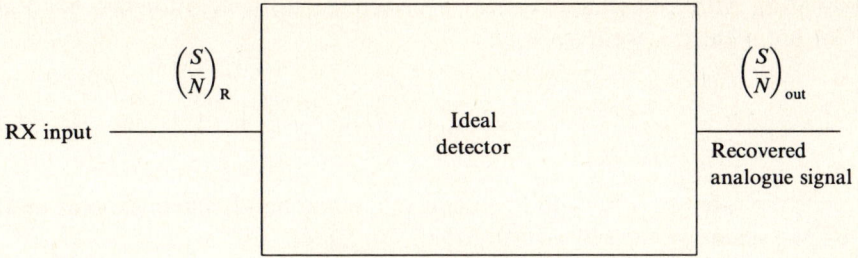

Figure 5.14   PCM receiver

Equation (5.70) based upon Shannon's channel capacity formula assumed that the noise component in $(S/N)_R$ is Gaussian. Although some Gaussian noise is introduced in transmission, as we saw earlier, noise in a PCM system is predominantly quantisation noise. Therefore, Eqn (5.70) is something of an approximation.

If the signal to noise ratios are large, Eqn (5.70) may be approximated:

$$B_T \log_2(S/N)_R = W \log_2(S/N)_{\text{out}} \tag{5.71}$$

If, in Eqn (5.71), antilogs are taken:

$$(S/N)_{\text{out}}^W = (S/N)_R^{B_T} \tag{5.72}$$

and if the Wth root is taken:

$$(S/N)_{\text{out}} = (S/N)_R^{B_T/W} \tag{5.73}$$

or, in decibel form:

$$(S/N)_{\text{out}}\, \text{dB} = \frac{B_T}{W} 10 \log_{10} \left(\frac{S}{N} \right)_R \tag{5.74}$$

B_T/W may be termed the bandwidth expansion factor and it is clear from Eqn (5.74) that the recovered analogue's signal to noise ratio in decibel is increased by this factor compared with the received signal. The analysis is based upon a single PCM channel but may readily be adapted for multiple channels.

We may make a crude comparison of $(S/N)_{\text{out}}$ for PCM compared with other forms of modulation. In Chapter 4 $(S/N)_{\text{in}}$ was used to represent the signal to noise ratio at the input to the demodulator which is the same point as the receiver input in Figure 5.14. In the case of PCM the noise bandwidth ηW of a baseband signal is increased by the bandwidth expansion factor. If, in comparing with a baseband system, received signal powers for baseband and PCM are identical, we may write:

$$(S/N)_{in} = (S/N)_R \frac{B_T}{W} \qquad (5.75)$$

That is the received S/N ratio of an equivalent baseband system exceeds that of a PCM system by the bandwidth expansion factor.

Rearranging Eqn (5.75) yields:

$$(S/N)_R = \frac{(S/N)_{in}}{B_T/W} \qquad (5.76)$$

We may develop a comparison between a PCM signal and an equivalent baseband digital signal in a similar way to Chapter 4, comparing modulated systems and equivalent baseband transmission.

Substituting Eqn (5.76) into Eqn (5.74) yields:

$$(S/N)_{out} \text{ dB} = \frac{B_T}{W} 10 \log_{10} \left[\frac{(S/N)_{in}}{B_T/W} \right] \qquad (5.77)$$

$$= \frac{B_T}{W} \left[(S/N)_{in} \text{ dB} - 10 \log_{10} \left(\frac{B_T}{W} \right) \text{ dB} \right] \qquad (5.78)$$

Equation (5.78) enables us to construct a graph, shown in Figure 5.15, of the performance of PCM against input signal to noise ratio for bandwidth expansion factors of 6, 9 and 13. $(S/N)_{in}$ is equivalent to $(S/N)_{out}$ of an equivalent baseband system, hence it is immediately apparent how $(S/N)_{out}$ of a PCM system is substantially greater than an equivalent baseband system. All three curves are remarkably similar, irrespective of bandwidth expansion factor. A very acceptable performance is possible from channels (and, indirectly, signals) having significant noise contributions. In addition only small improvements in the input signal to noise ratio lead to dramatic improvements in $(S/N)_{out}$.

If we compare the graph with similar curves for analogue modulation in Figure 4.23 we see that PCM offers a substantial improvement in signal to noise performance, even compared with FM. Thus PCM is a very effective bandwidth expansion-based system. Some care is necessary in comparing PCM with analogue systems. The earlier analysis leading to Eqn (5.78) makes reference to signal to noise ratios in a Gaussian sense. As we saw earlier in the chapter, noise in PCM predominantly arises from quantisation. Hence, to some extent, comparison of PCM and analogue modulation does not entirely compare like with like. Nevertheless, the general trends in terms of noise performance are correct.

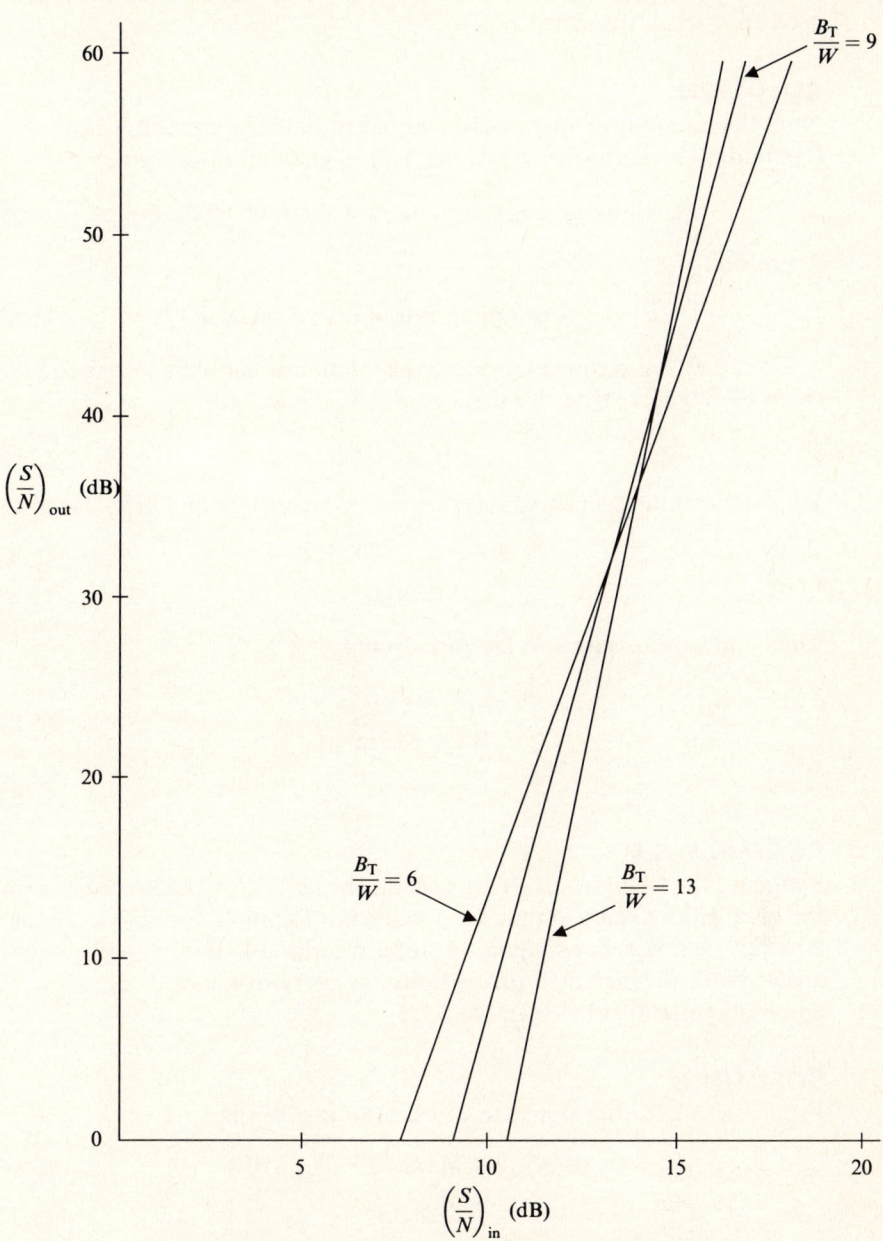

Figure 5.15   PCM transmission compared with baseband

EXAMPLE 5.4

Estimate the bandwidth expansion factor for an eight-bit PCM system used for speech transmission.

SOLUTION

Now the minimum value of B_T is equal to half the signalling rate (assuming a best case of 2 bits per Hz). Signalling rate is given by:

$$\text{Signalling rate} = \text{sampling rate} \times \text{no. of bits/sample} \quad (5.79)$$

Therefore:

$$B_T = \tfrac{1}{2} \times \text{sampling rate} \times \text{no. of bits/sample} \quad (5.80)$$

To satisfy the Nyquist criterion, the minimum sampling rate is equal to twice the highest signal frequency, or $2W$. Therefore:

$$B_T \geq NW \quad (5.81)$$

We shall assume speech to have an upper frequency of $4\,\text{kHz}$. Hence:

$$B_T = 8 \times 4000\,\text{Hz} \quad (5.82)$$

$$= 32\,\text{kHz} \quad (5.83)$$

The bandwidth expansion factor is found:

$$\frac{B_T}{W} = \frac{32\,\text{kHz}}{4\,\text{kHz}} \quad (5.84)$$

$$= 8 \quad (5.85)$$

EXAMPLE 5.5

Estimate, for an eight-bit PCM system, the $(S/N)_{\text{out}}$ of the receiver using the bandwidth expansion factor obtained in Example 5.4 above. Assume an equivalent baseband signal to noise ratio of $14\,\text{dB}$. Compare your answer with the signal to quantisation noise ratio for peak signal amplitude obtained in Example 5.3.1.

SOLUTION

From Eqn (5.78) the signal to noise ratio is given by:

$$(S/N)_{\text{out}} = 8(14 - 10\log_{10} 8)\,\text{dB} \quad (5.86)$$

$$= 39.8\,\text{dB} \quad (5.87)$$

Use of Eqn 5.78 to obtain $(S/N)_{\text{out}}$ for a PCM system is an alternative approach to that of the signal to quantisation noise approach earlier. In comparing results by each method we see that a $(S/N)_{\text{out}}$ of $39.8\,\text{dB}$ compares with a signal to quantisation noise of 52.8 dB in Example 5.2. Two possible reasons for the difference is that in Example 5.2, a peak signal amplitude was assumed which produces a best possible value of signal to quantisation noise ratio. The analysis used in this example to some extent takes account of noise contributed by the channel which the signal to quantisation noise approach ignores. This suggests that a more reasonable value of signal to noise ratio might lie between 52.8 and 39.8 dB.

We have already established that if the number of quantisation levels is increased, the signal to quantisation noise ratio is improved at the expense of a larger signalling rate and consummate increase in bandwidth. In examining the characteristics of a linear ADC, Figure 5.10, it is evident that for low input signal amplitudes, the ratio of signal to quantisation noise is less than for larger signals. Linear encoding requires that in order to provide adequate S/N at low signal levels an adequate number of quantisation levels is necessary. This leads to excessive S/N for high signal levels and an associated excessively large bandwidth for such signals.

It is possible to improve the quality of a PCM system without increasing the number of signalling levels and bandwidth required. That is by use of the complementary processes of compression and expansion, or **companding**.

If the ADC conversion process is made non-linear as, for example, in Figure 5.16 (shown only for positive signals), then a greater range of output binary values are assigned to smaller ranges of input voltage compared with larger

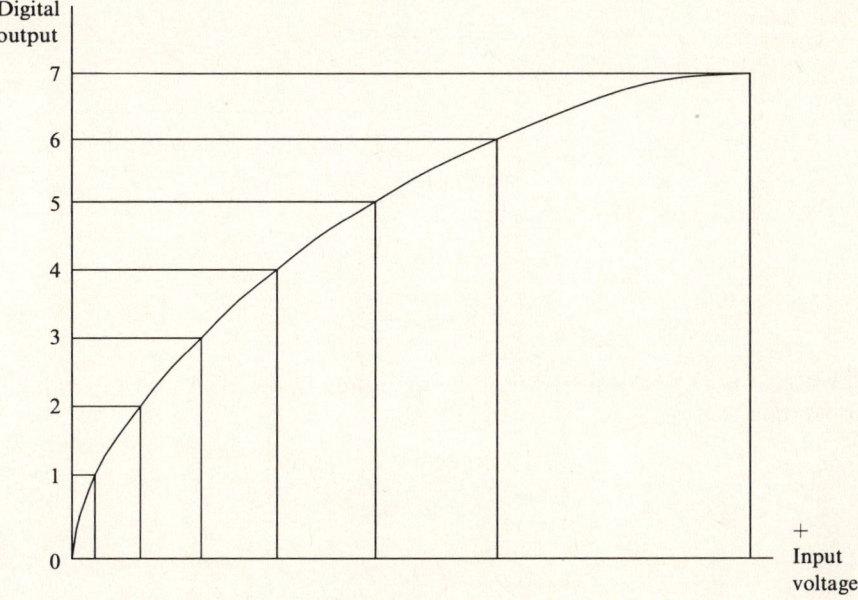

Figure 5.16 Compression

ones, a process known as **compression**. This means that the quantisation error for a given input voltage is lowest for small voltage and thereafter progressively increases. Ideally the aim is to provide constant signal to quantisation noise over the whole range of encoder input voltage. This is achieved by reducing quantisation error for small input voltages at the expense of increasing error for large voltages. In the latter case, as already suggested, increased quantisation noise occurs at larger input signal levels. Therefore signal to quantisation noise may well be comparable, dependent upon the non-linear characteristic employed, with that for smaller input signals. Thus the effect of quantising error is to produce a more uniform signal to quantisation noise, of satisfactory magnitude, over the whole range of input voltage.

At the receiver the encoder has a complementary characteristic to that of the encoder. In this way the overall effect of encoder and decoder in combination is to produce a linear response in that the recovered analogue signal is equal to that which is input at the transmitter, apart from any associated quantisation error.

Two slightly different encoding laws, both logarithmic in nature, have been established. In Europe the International Telecommunications Union–Telecomunications Sector (ITU–T) recommends:

$$v_{out} = \frac{1 + \log A v_{in}}{1 + \log A} \tag{5.88}$$

for:

$$\frac{1}{A} \preceq v_{in} \preceq 1$$

or:

$$v_{out} = \frac{A v_{in}}{1 + \log A} \tag{5.89}$$

for:

$$0 \preceq v_{in} \prec \frac{1}{A}$$

and where A is 87.6, so-called A-law companding. In the USA and Japan μ-law companding is used:

$$v_{out} = \frac{\log(1 + \mu v_{in})}{\log(1 + \mu)} \tag{5.90}$$

for:

$$0 \preceq v_{in} \preceq 1$$

(a)

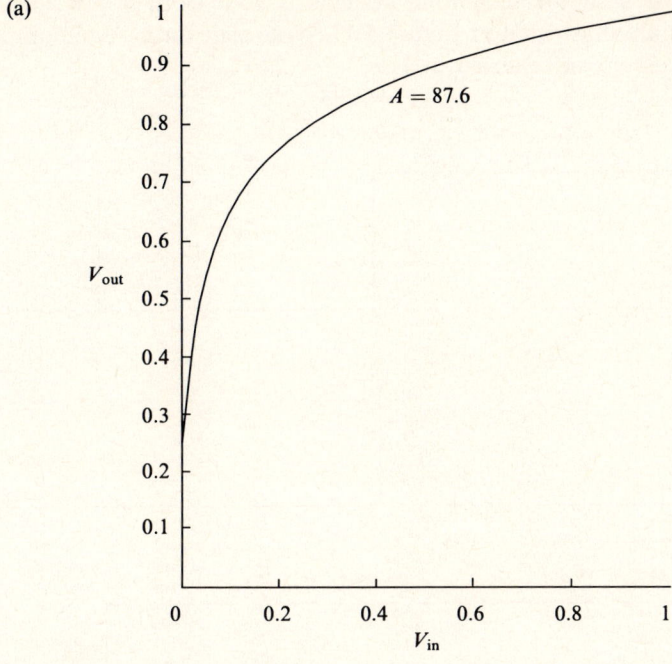

(b)

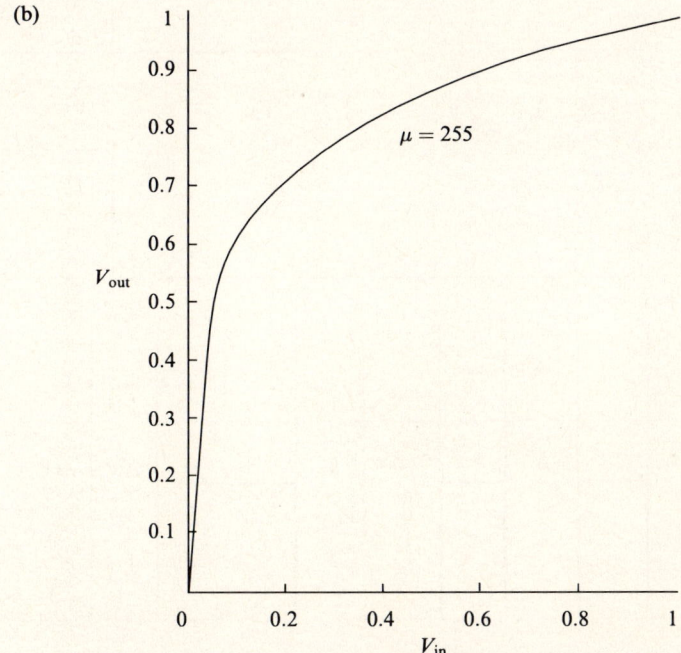

Figure 5.17 Compression characteristics of (a) A-law and (b) μ-law

The value of μ used for telephone systems is 255. Both A-law and μ-law characteristics are illustrated in Figure 5.17. Note that for ease of comparison both sets of axes are normalised to 1.

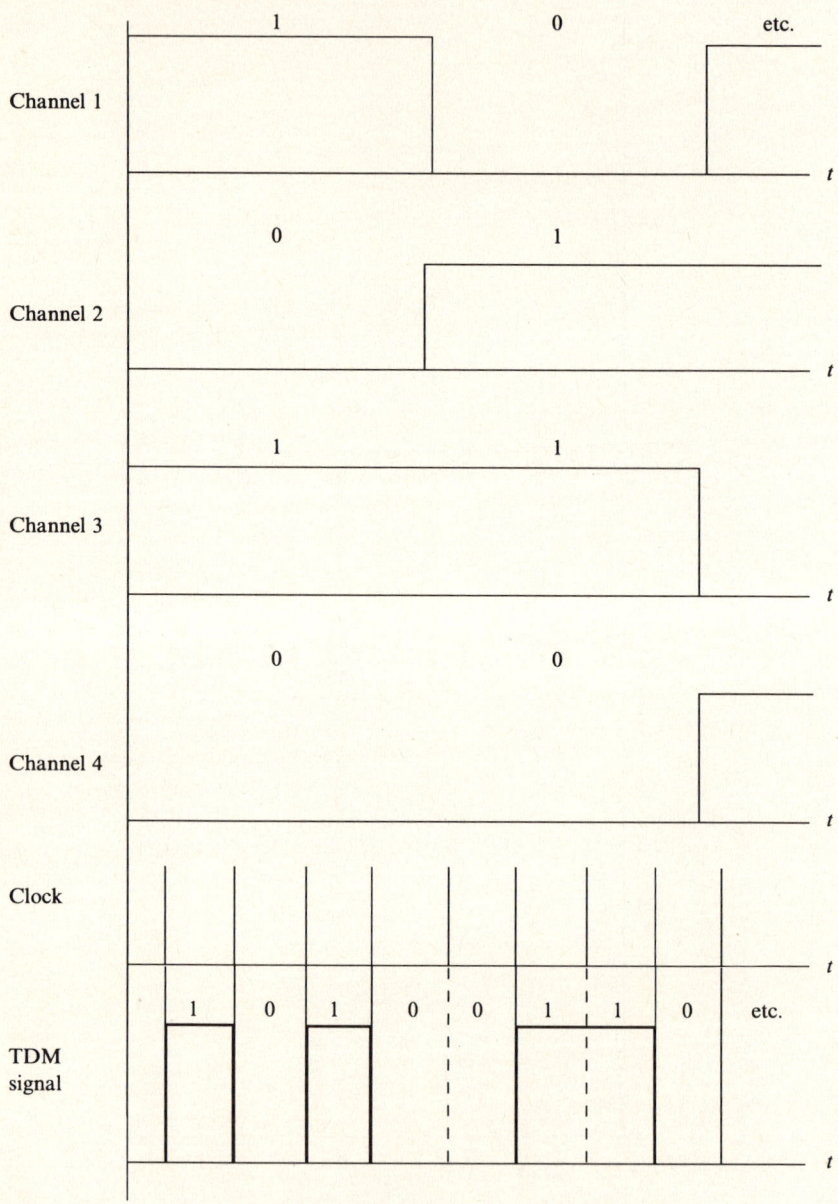

Figure 5.18   Multiplexing of four PCM signals

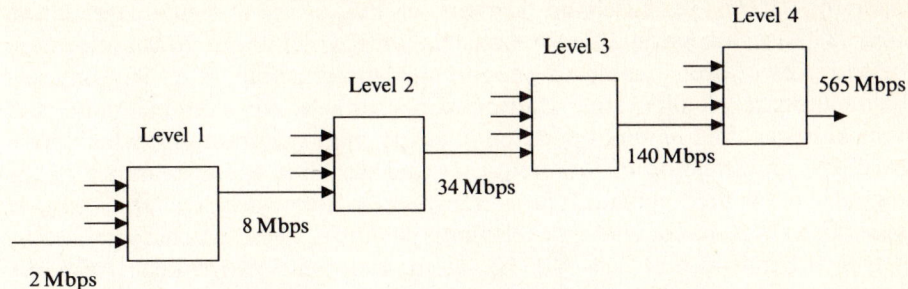

Figure 5.19 Plesiochronous digital trunk hierarchy

Before leaving PCM some mention of further development is necessary. As mentioned, PCM was originally developed for short-haul links between telephone exchanges. However, it was not long before PCM systems, although TDM systems in their own right, became further multiplexed. Four PCM systems may be multiplexed together as shown in Figure 5.18. Each system may simply be regarded as one 2 Mbps channel. The resultant TDM signal has a transmission rate of 8 Mbps.

The advantage of multiplexing a number of PCM systems was initially to make efficient use of existing coaxial cable systems originally laid between cities for trunk telephony traffic using frequency division multiplexing. Their conversion from analogue to digital operation led to cost reductions and opportunity to exploit the improved transmission quality of digital operation. Four PCM channels multiplexed together produce level 1, the primary level, of what is known as the **plesiochronous digital hierarchy** (PDH). Further multiplexing may be performed that further utilises the bandwidth available of coaxial cable or even higher bandwidth of optical fibre links. Figure 5.19 shows how at level 2, four 8 Mps streams can form a 34 Mbps stream, and so on, until a 565 Mbps stream may be produced. Signals at this speed may only be transmitted over optical links. Two such streams multiplexed together are now able to be carried commercially at terabit operation over fibre. Closer inspection of the resulting transmission rates after multiplexing indicates that some additional bits appear at the multiplexer output compared with that expected from the input streams. Some bits are added within multiplexers for housekeeping purposes, and in particular to assist in the synchronism of receivers to ensure that they demultiplex the incoming bit stream correctly.

5.4 Digital modulation

Baseband transmission of data generally requires a channel with a frequency spectrum from 0 Hz (dc) to a frequency equivalent to some multiple of the signalling rate. DC circuits have always been available from public telephone operators (PTOs) for baseband transmission but, owing to signal degradation over distance, are unsuitable for transmission beyond about 20 km and, until recently, have been unable to support a high signalling rate. The physical characteristics of alternative transmission channels, for example radio and optical links, often require operation in a particular spectral range which often precludes transmission of low frequency and dc. Until the appearance of the digital age, the predominant type of transmission link widely available to lower speed users was that of analogue telephone channels. These are characterised by a typical bandwidth of 300–3400 Hz, or so, and their low-frequency response precludes the transmission of baseband data.

To overcome distance limitations of dc circuits, data may be modulated on to a carrier which may be readily transmitted over an analogue channel. In this section we shall concentrate upon the use of analogue telephone channels since, even today, this remains a very common technique and is widely available throughout the world. In this way data can be communicated, via the dial-up telephone network, anywhere across the world using a telephone line. To this end **modems** (**mo**dulator/**dem**odulator) have been developed.

A sinusoidal carrier is described by its amplitude, frequency and phase, and each of these parameters may, as we saw in Chapter 4, be modulated or **keyed** by a modulating signal which is digital in nature. The term keyed has its roots in telegraphy where signalling used to be performed manually by means of a Morse-like key sender. The three basic forms of modulation, AM, FM and PM, are, in the case of digital modulation, known as **amplitude shift keying**, **frequency shift keying** or **phase shift keying** (ASK, FSK, PSK).

5.4.1 Amplitude shift keying

Modulation is a multiplying operation and it is assumed here that the carrier is sinusoidal with amplitude V and carrier frequency f_c. ASK is usually expressed in the time domain as a DSBSC signal:

$$g(t) = m(t) \sin 2\pi f_c t \tag{5.91}$$

The modulated signal's spectrum may be found by means of the Fourier transform which produces a pair of delta functions at carrier frequency, and equally spaced about 0 Hz, convolved with the Fourier transform of $m(t)$. We do not know precisely the time domain representation of $m(t)$. Hence we may only partially express $G(f)$ as shown:

$$G(f) = \mathcal{F}m(t) * \tfrac{1}{2}[\delta(f+f_c) + \delta(f-f_c)] \tag{5.92}$$

The effect of convolving a pair of delta functions with $M(f)$, the Fourier transform of $m(t)$, is to smear its spectrum about the frequencies $\pm f_c$. In practice $m(t)$ cannot be precisely specified because the pattern of 1s and 0s depends on the nature of the particular signal being transmitted. Consider the case of a repeated 1010 binary signal. From the Fourier series section in Chapter 2 we know that the spectrum of such a signal contains an infinite number of odd harmonics. However, adequate pulse shape of $m(t)$ is retained if the bandwidth is constrained such that only the fundamental and third harmonic are accepted. The time and frequency domain representations of such a signal are illustrated in Figure 5.20. It is a special case of ASK, where one symbol results in no transmission, known as **on–off keying** (OOK). The spectrum, as may have been expected from Eqn 5.91, contains both upper and lower sidebands and no carrier. We may deduce from the spectrum that a practical bandwidth of approximately six times the modulation rate, or three times the signalling rate R, is necessary.

We saw in Section 5.1.1 earlier that to reduce ISI, more sophisticated signal shaping than simple band-limiting, such as raised cosine pulse shaping, may be

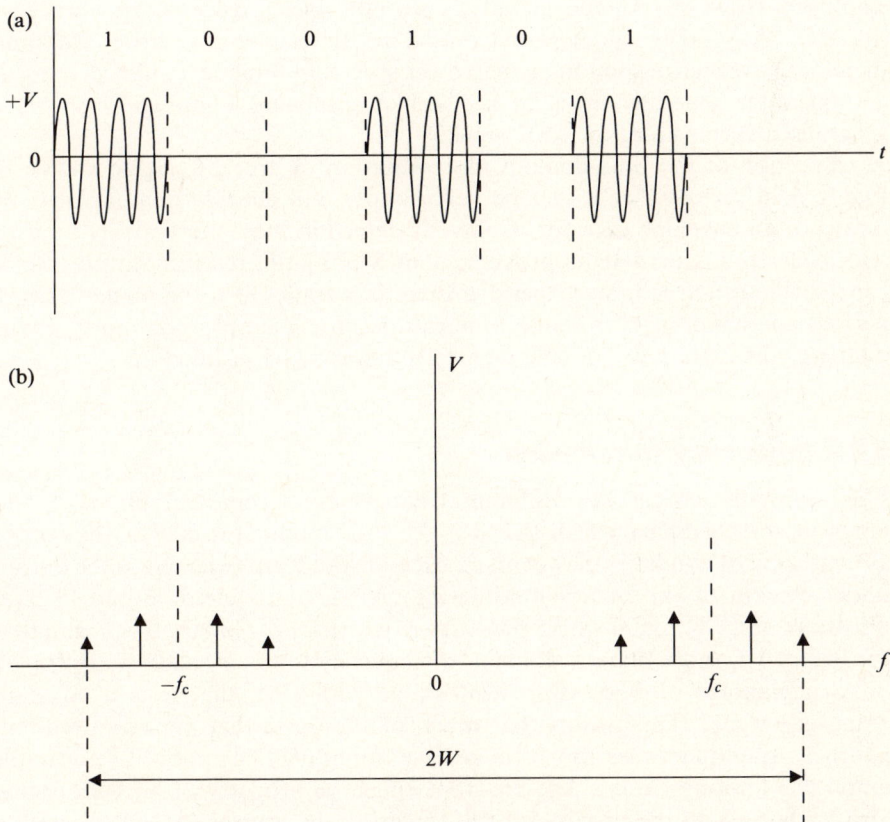

Figure 5.20   Amplitude shift keying: (a) time; (b) frequency

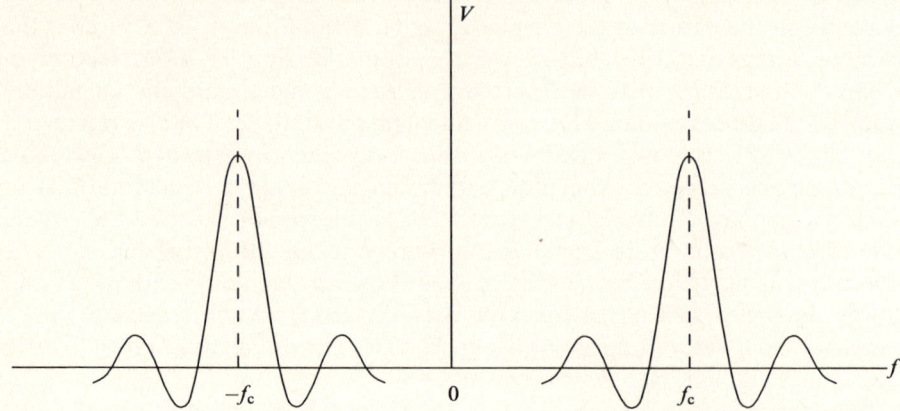

Figure 5.21 Generalised spectrum of an ASK signal

employed. However, the above analysis is useful since it represents a worst case situation. Any other signal would modulate the carrier at a lower rate and therefore have a correspondingly narrower spectrum. Practical pulse shaping at a transmitter generally leads to a sinc-like spectrum. Figure 5.21 shows a generalised spectrum of an ASK signal.

Inspection of the time domain representation of an ASK signal shown in Figure 5.20 reveals that it can be very simply and cheaply demodulated by means of an envelope detector. Coherent detection may alternatively be used and leads to a 3 or 4 dB improvement in S/N at the receiver output. Since coherent detection is more expensive there is a trade-off to be made between performance and cost. In some applications, for example very noisy environments, the extra costs of coherent detectors may be justified.

5.4.2 Frequency shift keying

FSK normally assigns two different frequencies to represent binary 1 and binary 0, so-called **binary FSK (BFSK)**. An FSK modulator may be thought of as consisting of two frequency sources, each of which are switched to the output under control of the binary modulating signal, as required. Figure 5.22(a) illustrates an FSK signal where the carrier frequency for binary 0 is f_0 and that for binary 1 f_1 (note that assignment of frequency to binary state is arbitrary).

An advantage of FSK over ASK is, as with FM, that it is a constant amplitude signal. This means that much of the noise that appears predominantly as amplitude variation at the receiver input may be removed by a simple amplitude limiting circuit. The constant envelope property of an FSK signal also lends itself to the provision of automatic gain control (AGC) to combat variation in signal level during transmission.

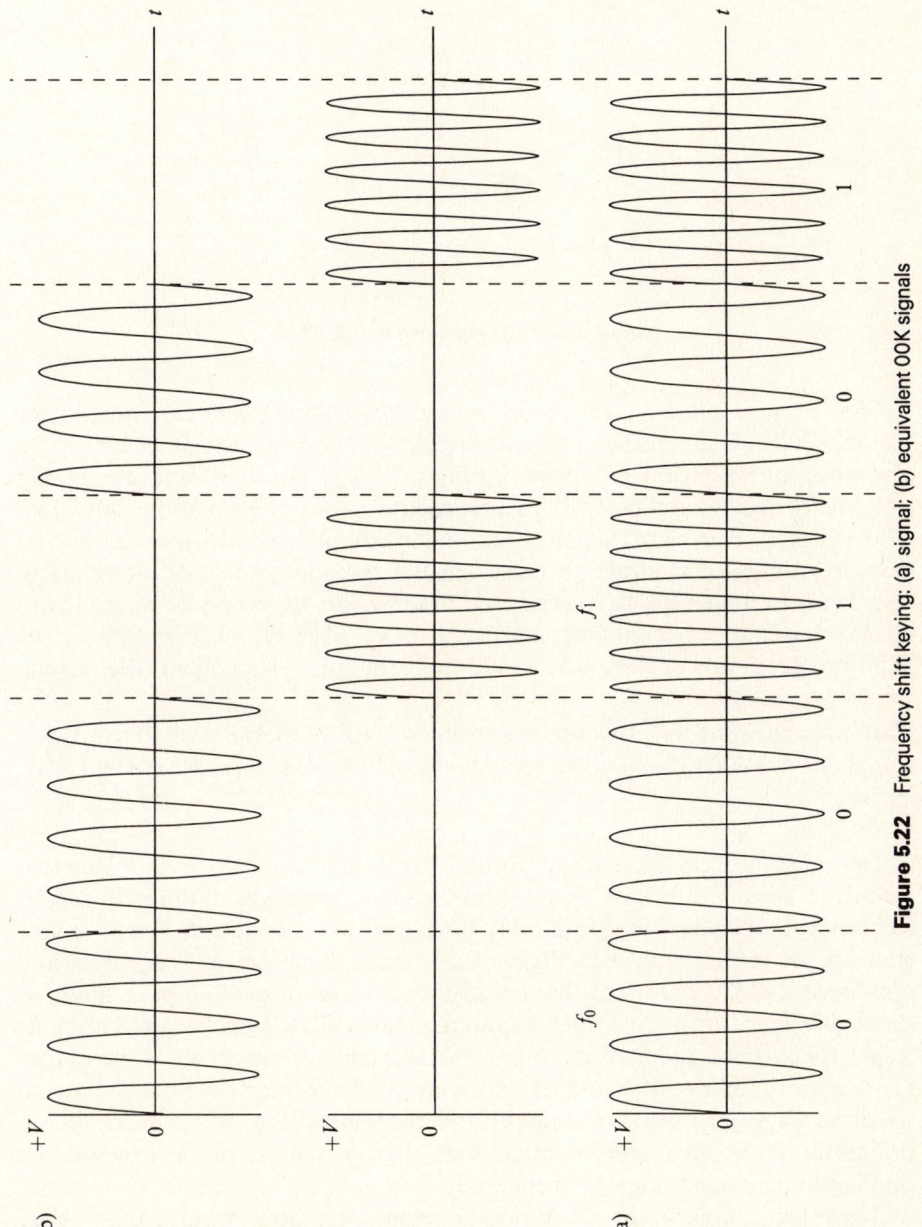

Figure 5.22   Frequency shift keying: (a) signal; (b) equivalent OOK signals

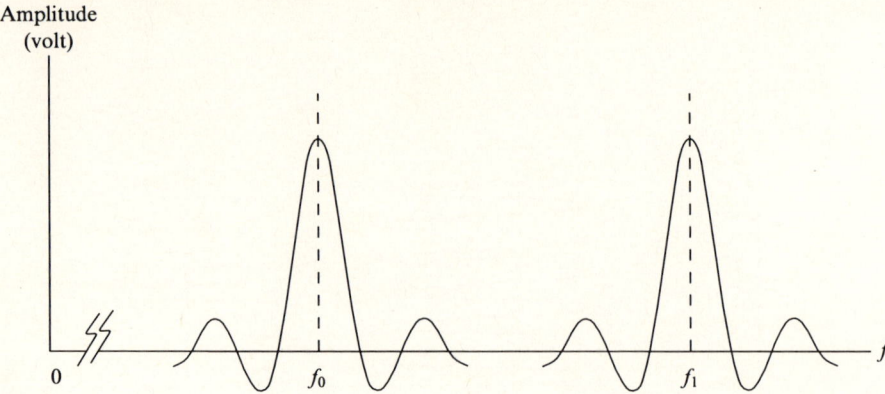

Figure 5.23 FSK spectrum (single sided)

FSK may be thought of as the sum of two separate OOK waveforms, Figure 5.22(b). One represents the occurrence of binary 1s, the other binary 0s. The resulting spectrum is shown in Figure 5.23. The two separate carrier frequencies must be sufficiently widely spaced to avoid overlap of the individual OOK spectra. This places a limit upon the maximum rate at which the carrier may be modulated in order to constrain the spreading effect of modulation. This tends to limit FSK to lower speed modulation, especially compared with phase modulation techniques discussed next. Clearly, in comparing, the bandwidth required of FSK is at least double that of an equivalent ASK signal.

Self-assessment 5.4 A modem is to employ BFSK modulation. If the data signal rate is 1000 bps, estimate the minimum bandwidth required. State any assumptions.

Detection of FSK may be performed by either coherent or non-coherent means. In considering FSK as two OOK signals, each with a different carrier frequency, it is evident that FSK may be detected non-coherently using simple envelope detection. Consider Figure 5.23 again. Each carrier frequency and associated OOK signal may be selected by means of a band-pass filter, as shown in Figure 5.24. Each filter output is a single OOK signal centred upon its respective carrier frequency. A binary 0 OOK signal produces a 1 at the output of its corresponding filter and envelope detector. Similarly a binary 1 signal results in a 1 at the output of its associated detector. If the output of the binary 0 detector is inverted and summed with that of the binary 1 detector, the original binary signal may be recovered.

There are a number of other non-coherent detection techniques for FSK which include frequency discrimination, zero-crossing detection and the use of a phase-locked loop. Improved noise performance may be achieved using

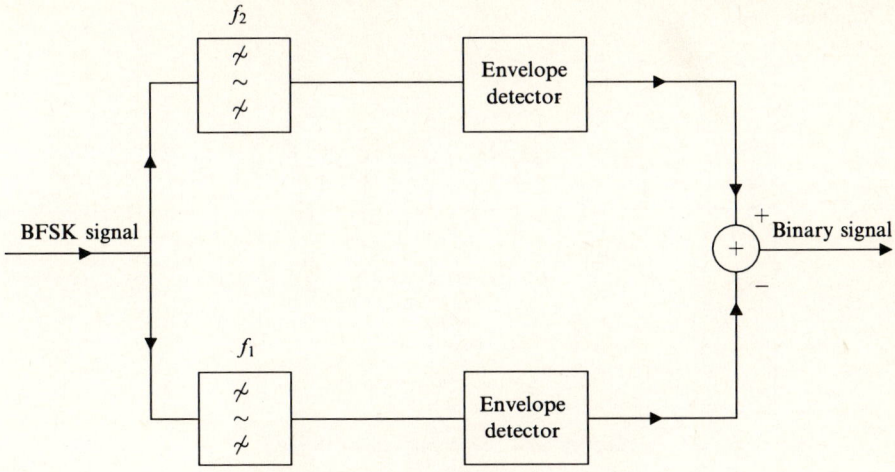

Figure 5.24 FSK reception (non-coherent)

coherent detection at additional cost and complexity. The complexity is exacerbated inasmuch that a BFSK signal comprises two carrier frequencies. Therefore two carrier recovery operations and demodulator stages are necessary, just as in non-coherent detection separate detectors are necessary for binary 1 and binary 0 signals, as we saw above. A coherent receiver is therefore similar to that shown in Figure 5.24 where envelope detectors are replaced by multipliers and additional carrier recovery circuits.

5.4.3 Phase shift keying

We shall initially consider binary phase shift keying, or **BPSK**, where a binary digital signal causes one of two phases to be transmitted, usually spaced π radian apart, as shown in Figure 5.25. In this example binary 1 causes transmission of zero phase and binary 0 transmission of π phase.

BPSK may be regarded as switching between two identical frequency sources of opposite phase. A BPSK signal may be represented mathematically:

$$g(t) = m(t)\sin 2\pi f_c t + [1 - m(t)](-\sin 2\pi f_c t) \qquad (5.93)$$

where, as before, $m(t)$ is the modulating signal and in the case of BPSK is either equal to 1 or 0 at any particular instant in time. Equation (5.93) may be rearranged:

$$g(t) = 2m(t)\sin 2\pi f_c t - \sin 2\pi f_c t \qquad (5.94)$$

If Eqn (5.94) is compared with Eqn (5.91), it may be deduced that PSK has a spectrum similar to that of OOK plus a component at carrier frequency. Such a spectrum is therefore similar to that of a full AM signal.

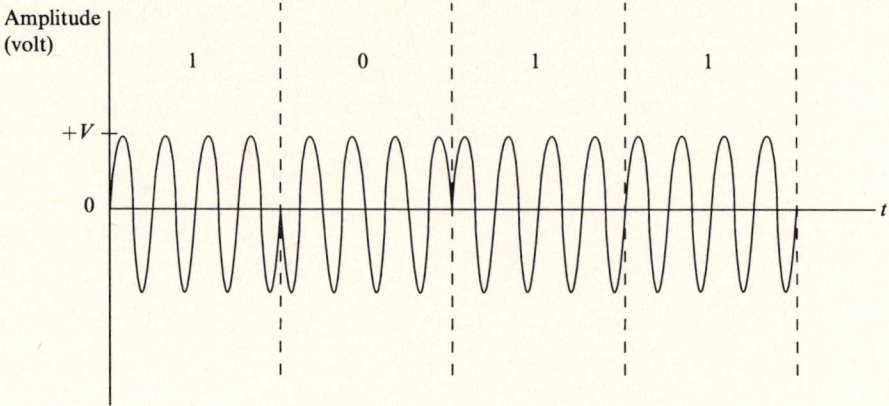

Figure 5.25   Binary phase shift keying

BPSK enjoys the advantages of FSK in that it is constant envelope, but also has the reduced bandwidth of ASK. BPSK is therefore spectrally efficient and, as we shall see later in subsection 5.4.5, performs better than both ASK and FSK in the presence of noise. The trade-off to be made when selecting BPSK is that, unlike ASK and FSK, it does not lend itself to non-coherent detection.

Differential phase shift keying (DPSK) may be employed to overcome the need for coherent detection necessary in PSK systems. In the simplest form of DPSK binary 1 may cause a phase shift of π whereas binary 0 causes no phase change, or vice versa. At the receiver the phase of each symbol is compared with that of the previous symbol, which means some method of delaying the received signal by one symbol length in time is necessary. A phase change indicates a 1 is received, no phase change a 0. Spectrally a binary modulated DPSK signal is similar to that of a BPSK signal.

A problem that may arise when using differential techniques is that, in this case, a long string of binary 0s would result in a long period of continuous phase transmission. The receiver can correctly detect the signal and output the appropriate binary state. However, absence of any symbol or recovered data signal boundaries means that the receiver is unable to retrieve a clock signal from the incoming signal. In general the receiver data clock would then drift from its optimum position. When a change in data state does occur at the receiver, some time may elapse before the clock is able to realign itself correctly. This happens only when there are sufficient transitions for the clock recovery circuit to re-synchronise reliably. During this time the data output from the receiver is incorrectly timed and may lead to errors in any following data equipment. To overcome this, modems commonly randomise data before application to a modulator at the transmitter using a scrambler. This ensures that no one symbol is transmitted continuously for any length of time.

All of the digital modulation methods, ASK, FSK, BPSK, discussed so far in this section are examples of binary systems. That is, the modulating signal only has two possible states resulting in modulation which only contains two symbols. However, there is in principle no reason why the modulator may not output more than two symbols. Such systems are called *m*-ary where *m* is the number of possible symbols. Consider the PSK case of four phases spaced $\pi/2$ apart. Such a system consisting of four symbols requires that binary data be taken two bits (dibits) at a time to produce one of four possible symbols. That is each symbol represents one of four dibits. Such a system is known as **quadrature PSK**, or QPSK. The more complex modulation strategies of *m*-ary systems are often shown in more convenient form as a **signal constellation**, for example Figure 5.26.

Figure 5.27(a) illustrates how the QPSK signal constellation of Figure 5.26 may be formed at a transmitter. The output of a two-bit shift register is latched every two bits. Bits b_1 and b_0 act as modulating signals to multiply their respective carriers at a symbol rate that is half that of the transmission rate of

Figure 5.26   QPSK signal constellation

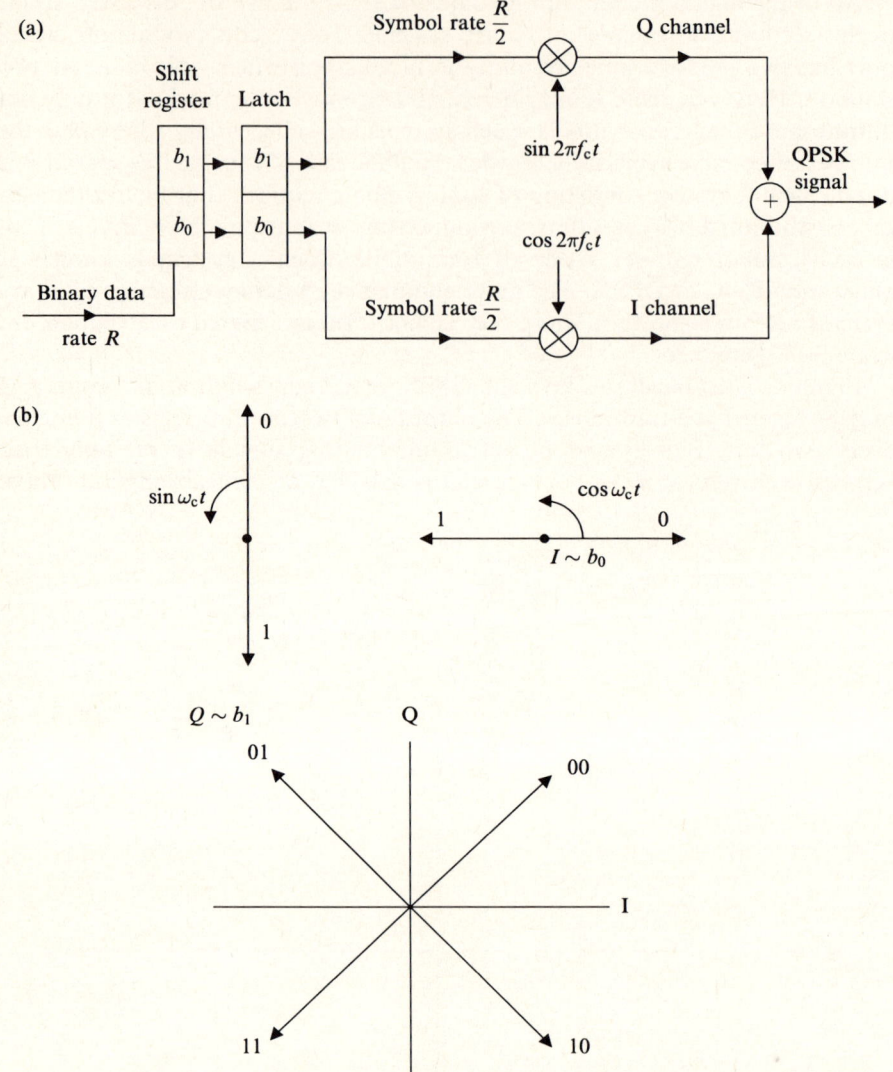

Figure 5.27 QPSK generation: (a) block diagram; (b) I and Q channels

the binary data. Hence the resultant QPSK signal also operates at half the transmission rate. The two carriers are in quadrature phase, hence the use of the term 'quadrature' in describing this form of modulation. The cosine carrier is regarded as the in-phase (or I) carrier and the sine term the quadrature (or Q) carrier. The I and Q signals form a pair of ASK, or BPSK, signals in phase quadrature. Figure 5.27(b) illustrates the signal constellation shown earlier in more detail. It is evident that the I and Q signals may be regarded as a BPSK signal. The I and Q signals are then summed to form a QPSK signal.

The use of quadrature carriers means that each BPSK signal is said to be **orthogonal**. A property of such signals is that they may coexist and yet not interfere with each other. In consequence, providing the channel does not disturb the phase of each BPSK signal such as to upset orthogonality, each BPSK signal contained within the received QPSK signal may be recovered using coherent techniques by means of two quadrature carriers.

The attraction of QPSK is that, because two BPSK spectra are summed, its bandwidth is equal to a single BPSK signal. This means that a channel of given bandwidth may transmit data using QPSK at twice the rate compared with BPSK. Alternatively, for the same transmission rate, QPSK operates at half the modulation rate of BPSK signalling, and hence the bandwidth may be halved. The above points serve to illustrate some of the advantages of m-ary systems. In practice some deterioration in BER results when BPSK and DPSK are compared for same data rate and signal amplitude. This is because points within a signal constellation of a QPSK signal are relatively closer together and therefore a smaller noise voltage disturbance is necessary to change a point from one symbol to that of another. QPSK, as with BPSK, may employ differential operation, obviating any need for carrier recovery at the receiver.

EXERCISE 5.6
Derive the relationship among data transmission rate D, symbol rate R and the number of symbols m for an m-ary system.

SOLUTION

$$\text{Transmission rate } D = R \times n \qquad (5.95)$$

where n is the number of bits per symbol. Now:

$$m = 2^n \qquad (5.96)$$

$$\therefore \quad n = \log_2 m \qquad (5.97)$$

Substituting Eqn (5.97) into (5.96):

$$D = R \times \log_2 m \qquad (5.98)$$

Clearly modulation of phase need not be constrained to a four point system. Since m must be integer, it follows from Eqn (5.96) that 8, 16, or more, phases may be employed. Irrespective of the size of m, a PSK signal constellation consists of points equally distributed around a circle (since all symbols have the same amplitude). For example, an eight phase system has eight points, each spaced $\pi/4$ apart. If the signal power is not increased with the number of points in the signal constellation, the amplitude of each symbol (radius of the circle upon which the points are positioned) also remains the same. This means that the spacing between points, especially those that are adjacent, progressively decreases with m. In consequence higher order PSK systems are more vulnerable to noise, distortion and interference and require greater resolution in distinguishing phase at the receiver.

Self-assessment 5.5
(a)   Sketch a suitable signal constellation for a 16-point PSK modulator. Clearly show how data bits may be mapped to the constellation assuming that orthogonal carriers are employed.
(b)   If the signalling rate is 4800 baud, determine the transmission rate in bit per second.

5.4.4 Quadrature amplitude modulation (QAM)

QAM, as its name suggests, also makes use of quadrature carriers. Unlike QPSK where each quadrature channel is modulated with a binary signal, QAM employs m-ary signalling on both I and Q channels. QAM is best illustrated by an example. Consider a 16-point QAM constellation. This is achieved by means of two four-point ASK signals as shown in Figure 5.28(a) by taking data two pairs at a time. That is dibits b_1 and b_0 modulate the in-phase carrier and b_3 and b_2 the quadrature carrier. Note that each dibit must be converted into one of four possible voltages for ASK modulation of its respective carrier. The resulting signal constellation is shown in Figure 5.28(b).

Self-assessment 5.6   Sketch I, Q and QAM signal constellations, in a similar manner to that shown for QPSK in Figure 5.27(b), for 16-point QAM and hence illustrate how Figure 5.28(b) results.

QAM uses signal space more efficiently than PSK by moving away from a circle which leads to greater spacing between adjacent signal points. This offers improved noise immunity without increase in system complexity. In practice a variety of signal constellation patterns are found rather than the rectangularly arranged pattern in Figure 5.28. Patterns seek to match the modulated signal to the phase and signal to noise characteristics of the channel to be employed. QAM is the sum of two orthogonal ASK signals. Hence its bandwidth equals that of a single ASK signal. Thus for a given ASK spectrum, two such ASK signals are contained within the same bandwidth.

Self-assessment 5.7   A modem employs QAM using a 64-point constellation. Sketch a suitable signal constellation. If the signalling rate is 1200 baud, determine the data transmission rate.

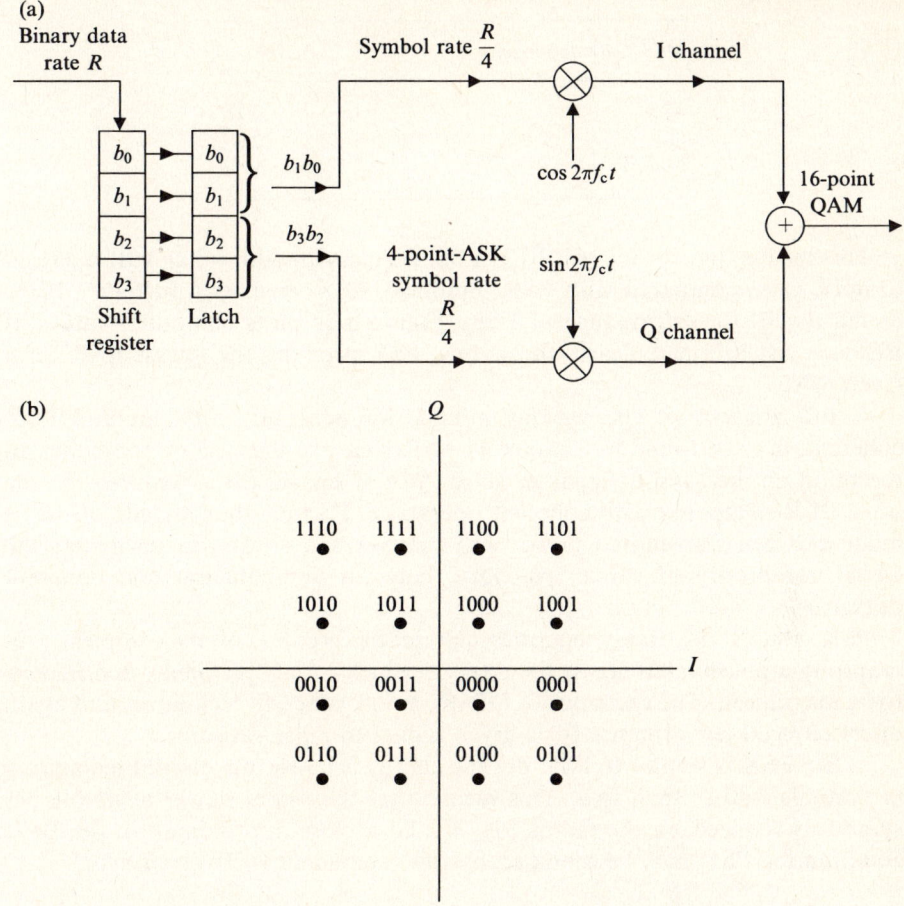

Figure 5.28   15-point QAM: (a) modulator; (b) signal constellation

5.4.5 Effect of noise upon digitally modulated signals

We saw earlier in the case of baseband digital signals that the occurrence of noise in transmission gives rise to the probability of an error at the receiver. Similarly, in transmitting digitally modulated signals, noise again gives rise to errors in detection at the receiver.

It may be shown that in the case of coherent detection that the BER is given by:

$$P_{\text{e}} = 0.5\left(1 - \text{erf}\sqrt{k\frac{S}{N}}\right) \tag{5.99}$$

where k is equal to that shown in Table 5.1.

Table 5.1 Comparison of coherent modulation

Modulation (coherent)	k
ASK	0.25
FSK	0.5
PSK	1

This means that for a given BER FSK requires an additional 3 dB in signal to noise ratio compared with PSK. Similarly ASK requires a further 3 dB to match the BER performance of FSK. Figure 5.29 plots computed values of BER against S/N and clearly shows how PSK provides the lowest BER for a given S/N.

A full analysis of the various modulation schemes, coherent and non-coherent, may be found in Chapter 11 of Roden[7]. In the case of non-coherent reception an increase in signal to noise ratio of several dB is required for the same BER compared with coherent operation. There is thus a trade-off to be made between transmitted power and receiver complexity, in particular the added complexity of carrier recovery, between non-coherent and coherent operation.

PSK almost invariably operates coherently because of its complexity in requiring a phase reference at the receiver which is only justified when BER is to be maximised. The exception is DPSK, a non-coherent technique, and again offers reduced performance for a given nignal to noise ratio.

M-ary PSK is similar to PSK but the energy is uniformly distributed among m symbols, rather than two. This means that the signal power available per symbol is reduced by the factor $\log_2 M$. To a close approximation the BER equation for PSK may be modified by this factor and BER given by:

$$P_e \approx 0.5\left[1 - \mathrm{erf}\sqrt{\left(\frac{S}{N}\frac{1}{\log_2 M}\right)}\right] \qquad (5.100)$$

That is for the same BER as for BPSK, S/N ratio is required to be increased by 3 dB per doubling of the number of symbols in the signal constellation.

QAM, as already mentioned, may be regarded as two coherent ASK signals, each transmitted on a quadrature carrier. In the case of a four-point signal constellation it is similar to a pair of binary ASK signals and therefore would require an additional 3 dB S/N ratio for similar BER performance. Where, as is more usual, M is larger than 4, improved BER performance becomes possible compared with equivalent m-ary PSK transmission. This is because for the same number of signal points separation between 'nearest neighbour' points is

---

[7] Roden, M.S., *Analog and Digital Communication Systems*, 4th edn, Prentice Hall, 1996. ISBN 0-13-399965-3.

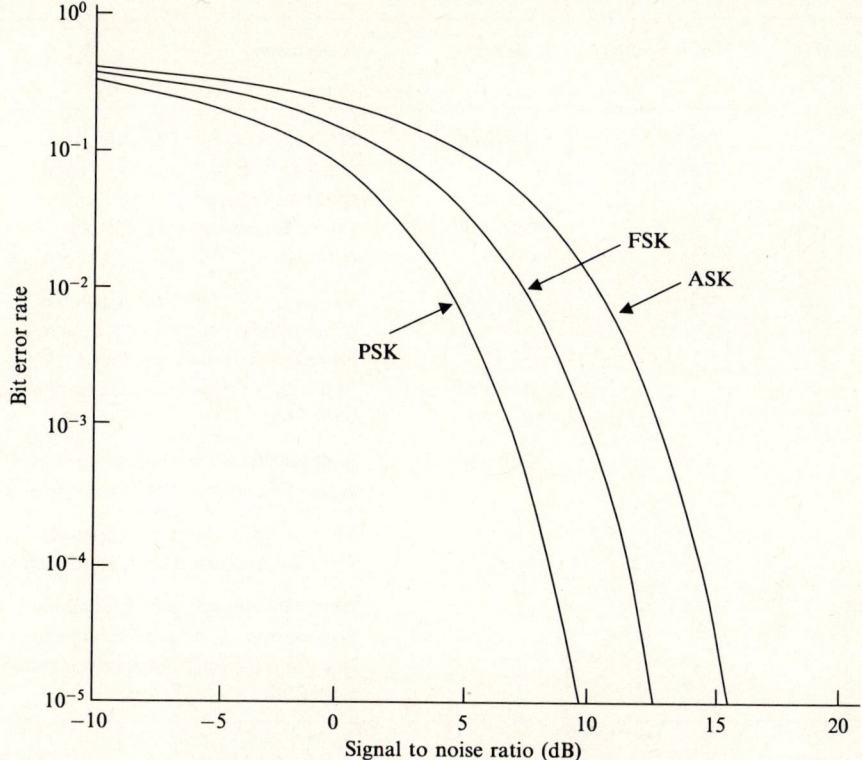

Figure 5.29 Bit error rate v. S/N (coherent detection)

increased in QAM compared with PSK. In consequence a larger noise amplitude must occur to cause a symbol (and its associated bits) to be misinterpreted at the receiver. As a result increasingly larger sized QAM constellations show increased BER performance compared with PSK.

5.4.6 Comparison of digital modulation techniques

Table 5.2 summarises the bandwidths required for various types of modulation and compares their BER performance for a given signal to noise ratio. As earlier, R is the signalling rate.

5.5 Summary

The Nyquist rate indicates that a digital signal requires a minimum bandwidth equal to half that of the signalling rate. Band-limiting of practical digital

Table 5.2 Comparison of modulation techniques

Modulation technique	Bandwidth	Detection	Performance
ASK	$\approx 3R$	Coherent	Similar to coherent PSK which has superior BER performance. Hence seldom employed.
		Non-coherent	Poor BER performance. Simple detector.
FSK	$> 3R$	Coherent	Similar to coherent PSK which has superior BER performance. Spectrally less efficient. Rarely employed.
		Non-coherent	Improved BER performance compared with ASK.
PSK	$\approx 3R$	Coherent	Improved BER performance compared with ASK and FSK. Spectrally efficient.
DPSK	$\approx 3R$		Less complex detector compared with PSK, for small loss in BER performance.
QAM	$\approx 3R$		More efficient signal constellations provide improved noise immunity. Improved BER performance compared with PSK.

signals at baseband causes spreading of the signal in time leading to inter-symbol interference. The effects of ISI may be examined qualitatively using an eye diagram. Equalisation may be used to modify the overall response of a channel to reduce the effect of ISI.

The addition of noise to a digital signal gives rise to errors at a receiver which are quantified as a bit error rate. Error function may be used to calculate BER for given signal amplitude and Gaussian noise. Channel capacity indicates the maximum transmission rate possible over a channel for specified bandwidth and signal to noise ratio. TDM may be used to multiplex a number of digital signals to make efficient use of the available bandwidth of a channel.

PCM is widely used, initially to increase the number of telephone circuits carried by a conducting pair, and subsequently as primary level inputs to higher order TDM systems. PCM is based upon the process of quantisation which inherently produces a sampling error, known as quantisation noise, and leads to the concept of signal to quantisation noise ratio which is the principal measure of performance. PCM is an example of a bandwidth expansion system offering improved signal to noise ratio performance at the expense of increased bandwidth.

The use of non-linear encoding, or companding, attempts to retain similar overall signal to noise ratio performance for reduced bandwidth by means of compressing stronger signal levels. ASK, FSK, PSK and QAM are commonly used to modulate digital signals onto a carrier. This is usually to exploit the readily available telephone network to support computer communication, or to make use of radio or optical fibre channels which cannot directly transmit digital signals at baseband frequencies.

Exercises

5.1 A typical PSTN link over which FSK data signals are carried has a bandwidth of 3000 Hz and S/N ratio of 20 dB. Determine the channel capacity.

5.2 Consider transmission of telephone signals by means of PCM. Given that the volume of individual speakers varies, explain what advantage non-linear encoding offers.

5.3 A PCM system is required to handle high-quality audio signals in the frequency range 0 to 20 kHz. Compare the relative values of signal to quantisation noise ratio (S/N_q), bandwidth and bandwidth expansion factor relative to analogue transmission at baseband for 8-bit and 12-bit sampling.

5.4 Explain the difference between the units bit per second and baud.

5.5 A modem employs a 16-phase QAM signal constellation. If data are transmitted at a rate of 2400 bps, determine the signalling rate and the number of bits transmitted per symbol.

5.6 Explain what is meant by the following statement: QPSK may be regarded as two ASK signals, each modulating one of two quadrature carriers.

Bibliography

Betts, J.A., *Signal Processing, Modulation and Noise*, The English Universities Press, 1970. ISBN 0-340-09895-3.

Bylanski, P. and Ingram, D.G.W., *Digital Transmission Systems*, 2nd edn, Peter Peregrinus, 1980. ISBN 0-906048-42-7.

Duck, M., Bishop, P. and Read, R., *Data Communications for Engineers*, Addison-Wesley, 1996. ISBN 0-201-42788-5.

Glover, I.A. and Grant, P.M., *Digital Communications*, Prentice Hall, 1997. ISBN 0-13-565391-6.

Marshall, G.M., *Principles of Digital Communications*, McGraw-Hill, 1980. ISBN 0-07-084096-2.

Roden, M.S., *Analog and Digital Communication Systems*, 4th edn, Prentice Hall, 1996. ISBN 0-13-399965-3.

Schwartz, M., *Information Transmission, Modulation and Noise*, 4th edn, McGraw-Hill, 1990. ISBN 0-07-100931-0.

Stremler, F.G., *Introduction to Communications Systems*, 3rd edn, Addison-Wesley, 1990. ISBN 0-201-51651-0.

Zeimer, R.E., and Tranter, W.H., *Principles of Communications*, 4th edn, John Wiley & Sons, 1995. ISBN 0-471-12496-6.

Solutions

Chapter 2

Self-assessment

2.1

$$g(t) = \frac{V}{2} + \frac{2V}{\pi}\left(\sin\frac{2\pi t}{T} + \frac{1}{3}\sin\frac{6\pi t}{T} + \frac{1}{5}\sin\frac{10\pi t}{T} + \cdots\right)$$

2.2

$$g(t) = \frac{V}{2} + \frac{V}{\pi}\left(\sin\frac{\pi t}{20} - \frac{1}{2}\sin\frac{\pi t}{10} + \frac{1}{3}\sin\frac{3\pi t}{20} - \frac{1}{4}\sin\frac{\pi t}{4} + \cdots\right)$$

2.3 $\operatorname{sinc} f\, e^{-j2\pi f t_d}$

2.4 $T = 0.5$

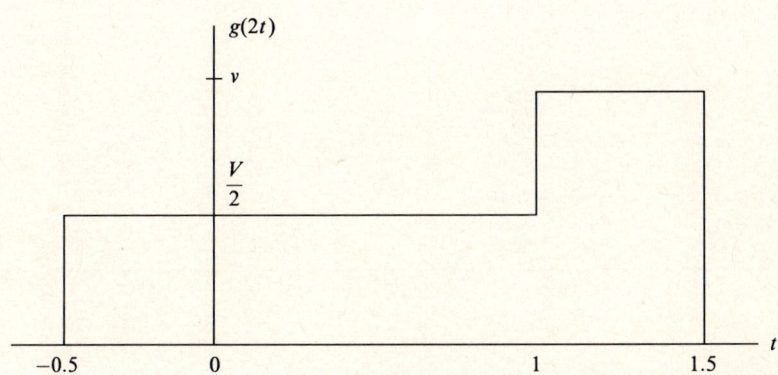

2.5   (a)

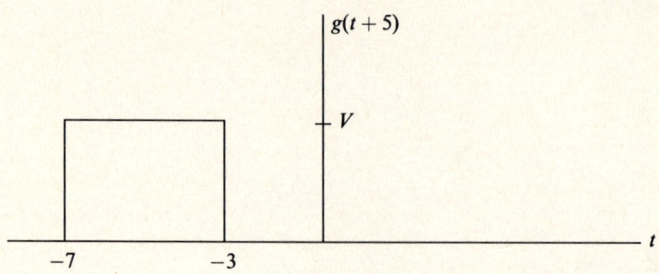

(b) $4V \operatorname{sinc} 4\pi f e^{j10\pi f}$

2.6 $20V \operatorname{sinc} 20\pi f e^{-j6\pi f}$

2.7

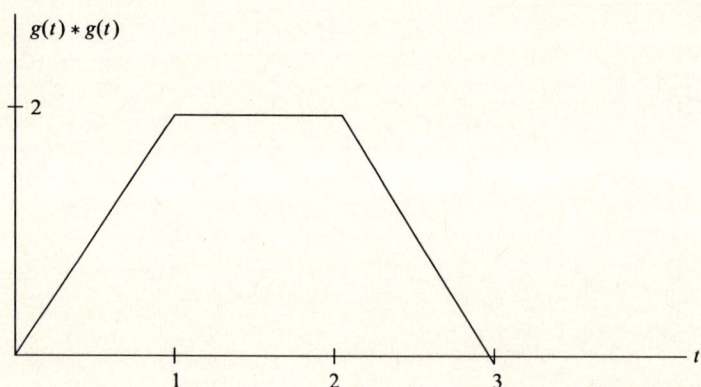

2.8

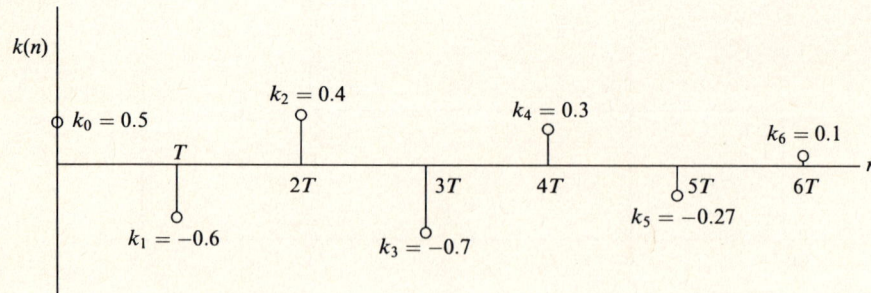

2.9

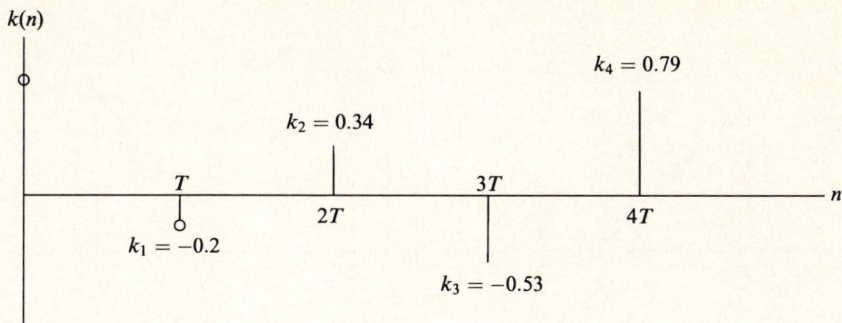

Exercises

2.1 (a) (i) Even; (ii) no symmetry; (iii) Even.

(b) (i) $\dfrac{V}{2} + \dfrac{4V}{\pi^2}\left(\cos\dfrac{2\pi t}{T} + \dfrac{1}{3^2}\cos\dfrac{6\pi t}{T} + \dfrac{1}{5^2}\cos\dfrac{10\pi t}{T} + \cdots\right)$

(ii) $\dfrac{V}{4} - \dfrac{2V}{\pi^2}\left(\cos\dfrac{2\pi t}{T} + \dfrac{1}{9}\cos\dfrac{6\pi t}{T} + \dfrac{1}{25}\cos\dfrac{10\pi t}{T} + \cdots\right)$

$$+\left(\sin\dfrac{2\pi t}{T} - \dfrac{1}{2}\sin\dfrac{4\pi t}{T} - \dfrac{1}{4}\sin\dfrac{8\pi t}{T} + \cdots\right)$$

(iii) $\dfrac{V}{\pi}\left(1 + \dfrac{\pi}{2}\cos\dfrac{2\pi t}{T} - \dfrac{2}{1.3}\cos\dfrac{4\pi t}{T} - \dfrac{2}{5.3}\cos\dfrac{8\pi t}{T}\right.$

$$\left.+\dfrac{2}{5.7}\cos\dfrac{12\pi t}{T} - \cdots\right)$$

2.3 $e^{-j2\pi fT}\operatorname{sinc} f + e^{j2\pi fT}\operatorname{sinc} f$

2.4 $\dfrac{2}{1 + (2\pi f)^2}$

2.5 (i) $\left|\dfrac{1}{F}\right| \cdot g(Ft)$

(ii) $\left[\dfrac{F}{2}\operatorname{sinc}\left(f\dfrac{F}{2}\right)\right]^2$

2.7

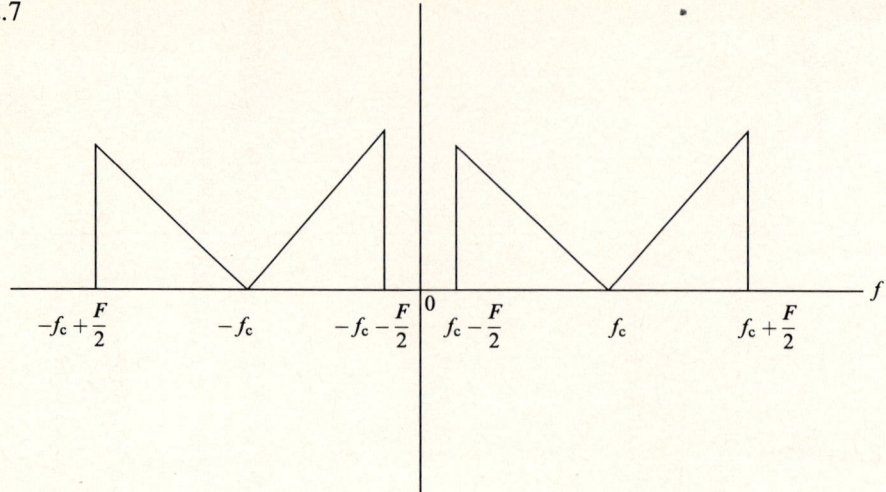

Chapter 3

Self-assessment

3.1 48.3 K.

3.2 (a) 900 K; (b) 0.033 pW; (c) 0.025 pW.

3.3 (a) 100; (b) 4 mW, 32 mW, 100 mW; (c) 0.4 W, 16 W.

3.4 (a) 4811 pW; (b) 4799 pW.

Exercises

3.1 $e_{TH} = \sqrt{[4kB(R_1 T_1 + R_2 T_2)]}$
     $R_{TH} = R_1 + R_2$

3.2 0.3 W, 0.0017 pW.

3.4 (a) 900 K; (b) 1.66 pW, 414 pW; (c) 601 nW.

3.5 Signal power 1 mW, 100 mW, 1.59 W. Noise power 1 mW, 15.9 mW, 100 mW.

3.6 (a) 379 pW; (b) 2.62 nW; (c) $\simeq 8$; (d) 1440°C.

3.8 (a) 1.014; (b) 120 K; (c) 1.058; (d) 36.82 K.

Chapter 4

Self-assessment

4.2 (a) 0.4; (b) 5 V; (c) 4 V.

4.3

m	Sideband : total power (%)	Carrier : total power (%)
0	0	100
0.1	≈ 0	99.5
0.2	2	98.0
0.3	4.3	95.7
0.4	7.4	92.6
0.5	11.1	88.9
0.6	15.3	84.7
0.7	19.7	80.3
0.8	24.2	75.8
0.9	28.8	71.2
1.0	33.3	66.7

4.4 9.1 nF.

Exercises

4.4 (a) 1; (b) 1000 Hz; (c) 0.33 μs; (d) 20 V; (e) 2 kHz.

4.5 0.32.

4.6 1.08 W.

4.7 303 W, 9697 W.

4.11 (a) 6.7; (b) 230 kHz; (c) 300 kHz $\prec W \prec$ 330 kHz.

Chapter 5

Self-assessment

5.1 0.9 V.

5.3 (b) 48.2 dB.

5.4   6 kHz.

5.5   (b) 19.2 kbps.

Exercises

5.1   19 963 bps.

5.3

	8 bit	12 bit	Analogue
S/N$_q$	$3M^2 = 52.9$ dB	89 dB	S/N 20–30 dB typ.
Bandwidth	160 kHz	240 kHz	20 kHz
Bandwidth expansion	8	2	–

5.5   600 baud, 4.

Fourier series

The series for a number of repetitive waveforms shown below are expressed in terms of ω where ω equals $2\pi/T$.

B.1 Square wave – even symmetry

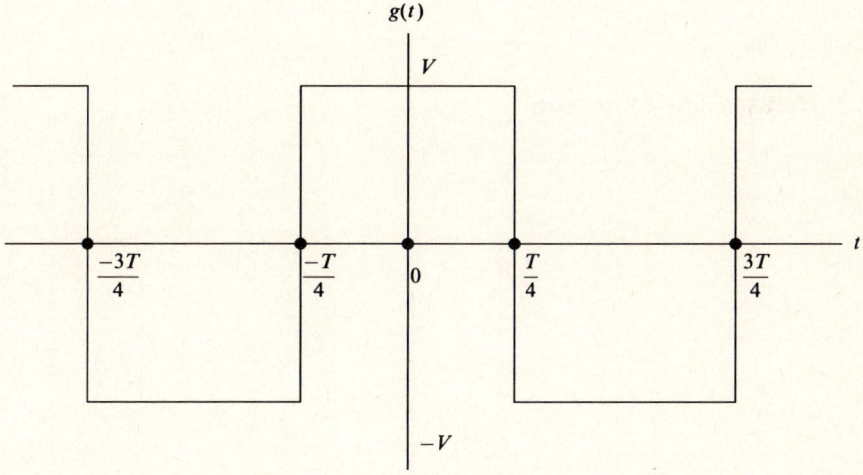

$$g(t) = \frac{4V}{\pi}(\cos \omega t - \tfrac{1}{3}\cos 3\omega t + \tfrac{1}{5}\cos 5\omega t - \cdots)$$

B.2 Square wave – odd symmetry

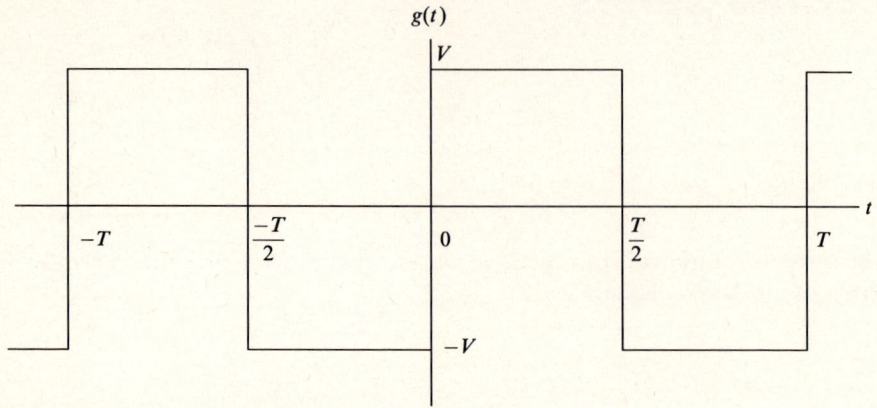

$$g(t) = \frac{4V}{\pi}(\sin \omega t + \tfrac{1}{3} \sin 3\omega t + \tfrac{1}{5} \sin 5\omega t + \cdots)$$

B.3 Rectangular wave

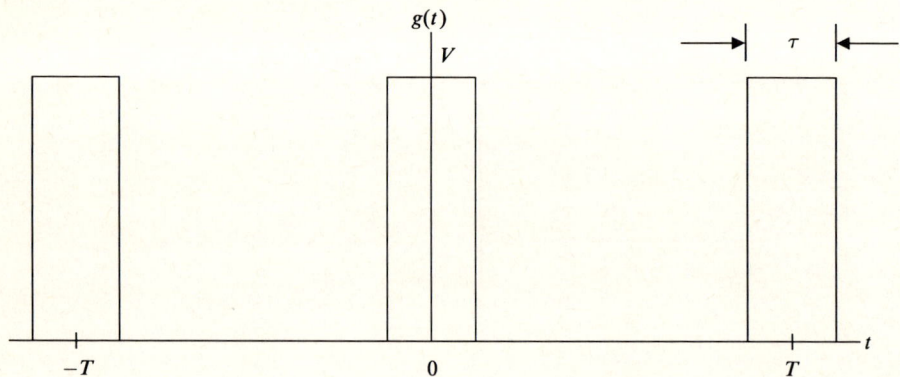

$$g(t) = \frac{2V}{\pi}\left[\frac{\tau}{4} + \sin\left(\frac{\tau}{2}\right) \cos \omega t + \tfrac{1}{2} \sin \tau \cos 2\omega t + \tfrac{1}{3} \sin\left(\frac{3\tau}{2}\right) \cos 3\omega t + \cdots\right]$$

B.4 Triangular wave

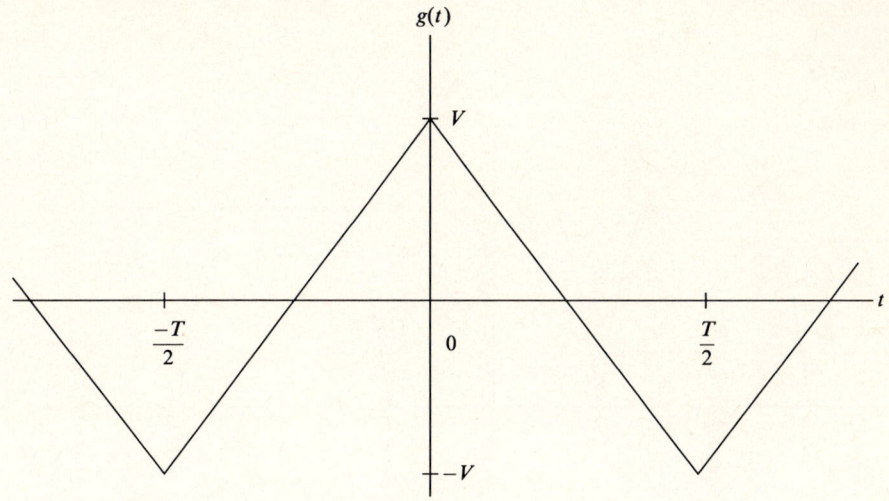

$$g(t) = \frac{8V}{\pi^2}\left(\cos \omega t + \frac{1}{3^2}\cos 3\omega t + \frac{1}{5^2}\cos 5\omega t + \cdots\right)$$

B.5 Sawtooth wave

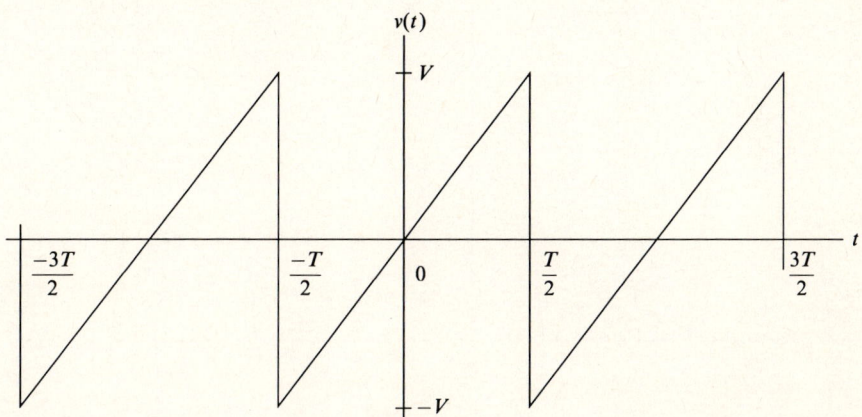

$$g(t) = \frac{2V}{\pi}(\sin \omega t - \tfrac{1}{2}\sin 2\omega t + \tfrac{1}{3}\sin 3\omega t - \tfrac{1}{4}\sin 4\omega t + \cdots)$$

B.6 Full-wave rectification

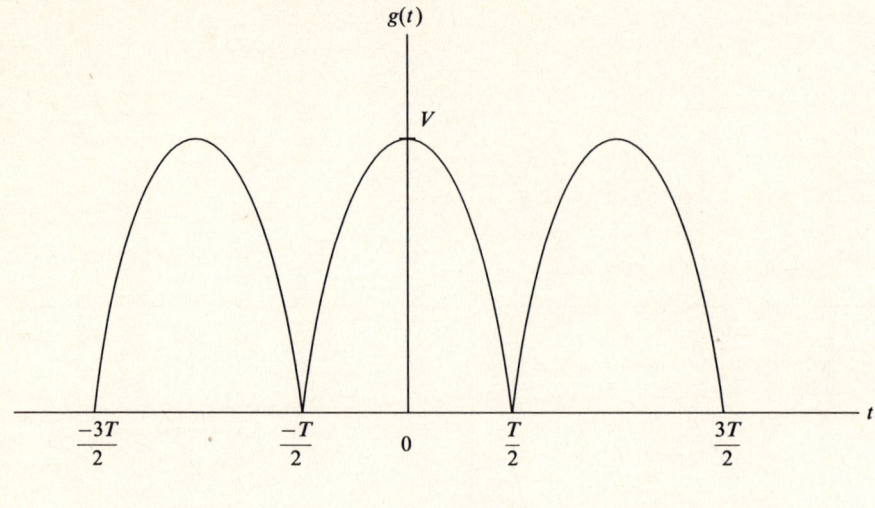

$$g(t) = \frac{2V}{\pi}\left(1 + \frac{2\cos 2\omega t}{1\times 3} - \frac{2\cos 4\omega t}{3\times 5} + \frac{2\cos 6\omega t}{5\times 7} - \cdots\right)$$

B.7 Half-wave rectification

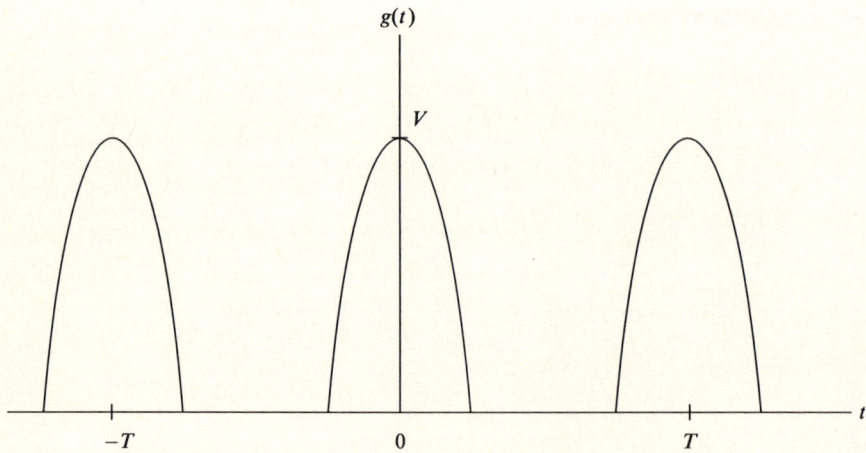

$$g(t) = \frac{V}{\pi}\left(1 + \frac{\pi}{2}\cos\omega t + \frac{2}{1\times 3}\cos 2\omega t - \frac{2}{5\times 3}\cos 4\omega t + \frac{2}{5\times 7}\cos 6\omega t - \cdots\right)$$

Fourier transform

C.1 General concepts

	$g(t)$	$G(f) = \int_{-\infty}^{\infty} g(t)\,\mathrm{e}^{-j2\pi ft}$
C.1.1 Scaling	$g(t/T)$	$\lvert T \rvert \cdot G(fT)$
C.1.2 Time shift	$g(t-T)$	$G(f)\,\mathrm{e}^{-j2\pi fT}$
C.1.3 Reciprocity	$G(t)$	$g(-f)$
C.1.4 Multiplication	$g(t) \cdot h(t)$	$G(f) * H(f)$
C.1.5 Convolution	$g(t) * h(t)$	$G(f) \cdot H(f)$

C.2 Selected functions

$g(t)$	$G(f) = \displaystyle\int_{-\infty}^{\infty} g(t)\,e^{-j2\pi ft}\,dt$

C.2.1 Delta function

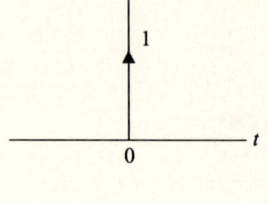

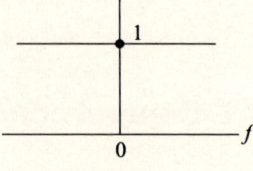

$$g(t) = \delta(t) \qquad\qquad G(f) = 1$$

C.2.2 Constant

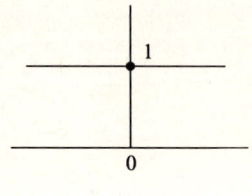

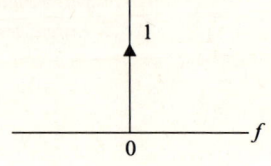

$$g(t) = 1 \qquad\qquad G(f) = \delta(f)$$

C.2.3 Rectangular function

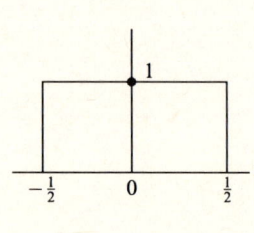

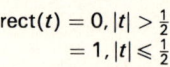

$$\operatorname{rect}(t) = 0,\ |t| > \tfrac{1}{2} \qquad\qquad \operatorname{sinc}(f) = \frac{\sin \pi f}{\pi f}$$
$$\phantom{\operatorname{rect}(t)} = 1,\ |t| \leqslant \tfrac{1}{2}$$

$g(t)$	$G(f) = \int_{-\infty}^{\infty} g(t)\, e^{-j2\pi ft}\, dt$

C.2.4 Unit step

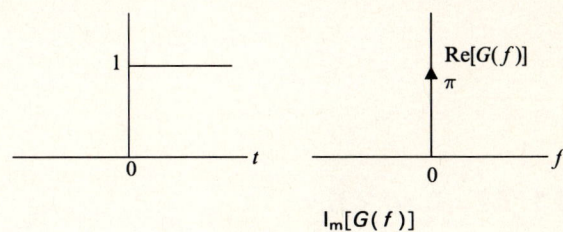

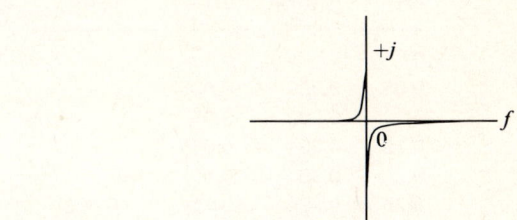

$$u(t) = 1, t < 0$$
$$\quad = 1, t \geqslant 0$$

$$G(f) = \tfrac{1}{2}\delta(f) - \frac{j}{2\pi f}$$

C.2.5 Exponential decay

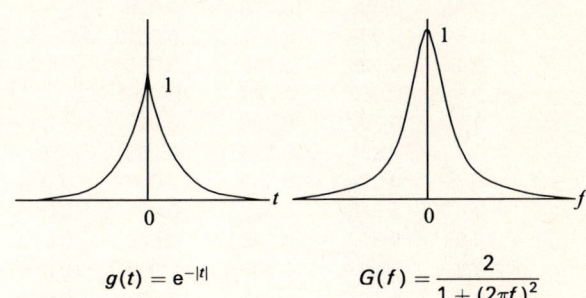

$$g(t) = e^{-|t|}$$

$$G(f) = \frac{2}{1 + (2\pi f)^2}$$

C.2.6 Repeated function

$$\text{rep}_T[g(t)] + g(t) * \text{rep}_T[\delta(t)]$$

$$G(f) \cdot \left|\frac{1}{T}\right| \text{rep}_{1/T}[\delta(f)]$$

C.2.7 Sampled function

$$g(t) \cdot \text{rep}_T[\delta(t)]$$

$$G(f) * \left|\frac{1}{T}\right| \text{rep}_{1/T}[\delta(f)]$$

APPENDIX D
Bessel functions

D.1 $M = 0$ to 2.5

M_1	$J_0(M)$	$J_1(M)$	$J_2(M)$	$J_3(M)$	$J_4(M)$	$J_5(M)$
0	1	0				
0.05	0.999	0.025				
0.1	0.998	0.05				
0.15	0.994	0.075				
0.2	0.99	0.1				
0.25	0.984	0.124	0.008			
0.3	0.978	0.148	0.011			
0.35	0.97	0.172	0.015			
0.4	0.96	0.196	0.02			
0.45	0.95	0.219	0.025			
0.5	0.938	0.242	0.031			
0.55	0.926	0.265	0.037			
0.6	0.912	0.287	0.044			
0.65	0.897	0.308	0.051			
0.7	0.881	0.329	0.059			
0.75	0.864	0.349	0.067	0.009		
0.8	0.846	0.369	0.076	0.01		
0.85	0.827	0.388	0.085	0.012		
0.9	0.808	0.406	0.095	0.014		
0.95	0.787	0.423	0.105	0.017		
1	0.765	0.44	0.115	0.02		
1.05	0.743	0.456	0.126	0.023		
1.1	0.72	0.471	0.137	0.026		
1.15	0.696	0.485	0.148	0.029		
1.2	0.671	0.498	0.159	0.033		
1.25	0.646	0.511	0.171	0.037		
1.3	0.62	0.522	0.183	0.041		
1.35	0.594	0.532	0.195	0.046		
1.4	0.567	0.542	0.207	0.05		
1.45	0.54	0.55	0.22	0.056	0.009	
1.5	0.512	0.558	0.232	0.061	0.01	
1.55	0.484	0.564	0.245	0.067	0.012	
1.6	0.455	0.57	0.257	0.073	0.013	

M_1	$J_0(M)$	$J_1(M)$	$J_2(M)$	$J_3(M)$	$J_4(M)$	$J_5(M)$
1.65	0.427	0.574	0.269	0.079	0.015	
1.7	0.398	0.578	0.282	0.085	0.017	
1.75	0.369	0.58	0.294	0.092	0.019	
1.8	0.34	0.582	0.306	0.099	0.021	
1.85	0.311	0.582	0.318	0.106	0.023	
1.9	0.282	0.581	0.33	0.113	0.026	
1.95	0.253	0.579	0.342	0.121	0.028	
2	0.224	0.577	0.353	0.129	0.031	
2.05	0.195	0.573	0.364	0.137	0.034	
2.1	0.167	0.568	0.375	0.145	0.037	
2.15	0.138	0.563	0.385	0.154	0.04	0.009
2.2	0.11	0.556	0.395	0.162	0.044	0.010
2.25	0 083	0.548	0.405	0.171	0.048	0.011
2.3	0.056	0.54	0.414	0.18	0.052	0.012
2.35	0.029	0.53	0.423	0.189	0.056	0.013
2.4	0.003	0.52	0.431	0.198	0.06	0.015
2.45		0.509	0.439	0.207	0.064	0.016
2.5					0.069	0.018

D.2 $M = 0$ to 20

M	$J_0(M)$	$J_1(M)$	$J_2(M)$	$J_3(M)$	$J_4(M)$	$J_5(M)$	$J_6(M)$
0	1						
1	0.765	0.44	0.115	0.02	0.003		
2	0.224	0.577	0.353	0.129	0.034	0.007	0.001
3	−0.26	0.339	0.486	0.309	0.132	0.043	0.011
4	−0.397	−0.066	0.364	0.43	0.281	0.132	0.049
5	−0.178	−0.328	0.047	0.365	0.391	0.261	0.131
6	0.151	−0.277	−0.243	0.115	0.358	0.362	0.246
7	0.3	−0.005	−0.301	−0.168	0.158	0.348	0.339
8	0.172	0.235	−0.113	−0.291	−0.105	0.186	0.338
9	−0.09	0.245	0.145	−0.181	−0.265	−0.055	0.204
10	−0.246	0.043	0.255	0.058	−0.22	−0.234	−0.014
11	−0.171	−0.177	0.139	0.227	−0.015	−0.238	−0.202
12	0.048	−0.223	−0.085	0.195	0.182	−0.073	−0.244
13	0.207	−0.07	−0.218	0.003	0.219	0.132	−0.118
14	0.171	0.133	−0.152	−0.177	0.076	0.22	0.081
15	−0.014	0.205	0.042	−0.194	−0.119	0.13	0.206
16	−0.175	0.09	0.186	−0.044	−0.203	−0.057	0.167
17	−0.17	−0.098	0.158	0.135	−0.111	−0.187	0.007
18	−0.013	−0.188	−0.008	0.186	0.07	−0.155	−0.156
19	0.147	−0.106	−0.158	0.072	0.181	0.004	−0.179
20	0.167	0.067	−0.16	−0.099	0.131	0.151	−0.055

M	$J_7(M)$	$J_8(M)$	$J_9(M)$	$J_{10}(M)$	$J_{11}(M)$	$J_{12}(M)$
1						
2						
3	0.003					
4	0.015	0.004				
5	0.053	0.018	0.006			
6	0.13	0.057	0.021	0.007		
7	0.234	0.128	0.059	0.024	0.008	0.003
8	0.321	0.223	0.126	0.061	0.026	0.01
9	0.327	0.305	0.215	0.125	0.062	0.027
10	0.217	0.318	0.292	0.207	0.123	0.063
11	0.018	0.225	0.309	0.28	0.201	0.122
12	−0.17	0.045	0.23	0.3	0.27	0.195
13	−0.241	−0.141	0.067	0.234	0.293	0.262
14	−0.151	−0.232	−0.114	0.085	0.236	0.285
15	0.034	−0.174	−0.22	−0.09	0.1	0.237
16	0.183	−0.007	−0.19	−0.206	−0.068	0.112
17	0.188	0.154	−0.043	−0.199	−0.191	−0.049
18	0.051	0.196	0.123	−0.073	−0.204	−0.176
19	−0.116	0.093	0.195	0.092	−0.098	0.205
20	−0.184	−0.074	0.125	0.186	0.061	−0.119

M	$J_{13}(M)$	$J_{14}(M)$	$J_{15}(M)$	$J_{16}(M)$	$J_{17}(M)$	$J_{18}(M)$
1						
2						
3						
4						
5						
6						
7						
8	0.003					
9	0.011	0.004				
10	0.029	0.012	0.005			
11	0.064	0.03	0.013	0.005		
12	0.12	0.065	0.032	0.014	0.006	
13	0.19	0.119	0.066	0.033	0.015	0.006
14	0.254	0.186	0.117	0.066	0.034	0.016
15	0.279	0.246	0.181	0.116	0.067	0.035
16	0.237	0.272	0.24	0.177	0.115	0.067
17	0.123	0.236	0.267	0.234	0.174	0.114
18	−0.031	0.132	0.236	0.261	0.229	0.171
19	0.161	0.015	0.139	0.234	0.256	0.224
20	−0.204	−0.146		0.145	0.233	0.251

M	$J_{19}(M)$	$J_{20}(M)$
1		
2		
3		
4		
5		
6		
7		
8		
9		
10		
11		
12		
13		
14	0.007	
15	0.017	0.007
16	0.035	0.017
17	0.067	0.036
18	0.113	0.067
19	0.168	0.112
20	0.219	0.165

Normal error function

$$\mathrm{erf}(x) = \frac{2}{\sqrt{\pi}} \int_0^{fx} \mathrm{e}^{-u^2} \, \mathrm{d}u$$

x	0	1	2	3	4	5	6	7	8	9
0	0	0.011 283	0.022 565	0.033 841	0.045 111	0.056 372	0.067 622	0.078 858	0.090 078	0.101 281
0.1	0.112 463	0.123 623	0.134 758	0.145 867	0.156 947	0.167 996	0.179 012	0.189 992	0.200 936	0.211 84
0.2	0.222 703	0.233 522	0.244 296	0.255 023	0.2657	0.276 326	0.2869	0.297 418	0.307 88	0.318 283
0.3	0.328 627	0.338 908	0.349 126	0.359 279	0.369 365	0.379 382	0.389 33	0.399 206	0.409 009	0.418 739
0.4	0.428 392	0.437 969	0.447 468	0.456 887	0.466 225	0.475 482	0.484 655	0.493 745	0.50275	0.511 668
0.5	0.5205	0.529 244	0.537 899	0.546 464	0.554 939	0.563 323	0.571 616	0.579 816	0.587 923	0.595 936
0.6	0.603 856	0.611 681	0.619 411	0.627 046	0.634 586	0.642 029	0.649 377	0.656 628	0.663 782	0.67084
0.7	0.677 801	0.684 666	0.691 433	0.698 104	0.704 678	0.711 156	0.717 537	0.723 822	0.73001	0.736 103
0.8	0.742 101	0.748 003	0.753 811	0.759 524	0.765 143	0.770 668	0.7761	0.781 44	0.786 687	0.791 843
0.9	0.796 908	0.801 883	0.806 768	0.811 564	0.816 271	0.820 891	0.825 424	0.82987	0.834 232	0.838 508
1	0.842 701	0.84681	0.850 838	0.854 784	0.85865	0.862 436	0.866 144	0.869 773	0.873 326	0.876 803
1.1	0.880 205	0.883 533	0.886 788	0.889 971	0.893 082	0.896 124	0.899 096	0.902	0.904 837	0.907 608
1.2	0.910 314	0.912 956	0.915 534	0.91805	0.920 505	0.9229	0.925 236	0.927 514	0.929 734	0.931 899
1.3	0.934 008	0.936 063	0.938 065	0.940 015	0.941 914	0.943 762	0.945 561	0.947 312	0.949 016	0.950 673
1.4	0.952 285	0.953 852	0.955 376	0.956 857	0.958 297	0.959 695	0.961 054	0.962 373	0.963 654	0.964 898
1.5	0.966 105	0.967 277	0.968 413	0.969 516	0.970 586	0.971 623	0.972 628	0.973 603	0.974 547	0.975 462
1.6	0.976 348	0.977 207	0.978 038	0.978 843	0.979 622	0.980 376	0.981 105	0.98181	0.982 493	0.983 153
1.7	0.98379	0.984 407	0.985 003	0.985 578	0.986 135	0.986 672	0.98719	0.987 691	0.988 174	0.988 641
1.8	0.989 091	0.989 525	0.989 943	0.990 347	0.990 736	0.991 111	0.991 472	0.991 821	0.992 156	0.992 479
1.9	0.99279	0.99309	0.993 378	0.993 656	0.993 923	0.994 179	0.994 426	0.994 664	0.994 892	0.995 111

x										
2	0.995322	0.995525	0.995719	0.995906	0.996086	0.996258	0.996423	0.996582	0.996734	0.99688
2.1	0.997021	0.997155	0.997284	0.997407	0.997525	0.997639	0.997747	0.997851	0.997951	0.998046
2.2	0.998137	0.998224	0.998308	0.998388	0.998464	0.998537	0.998607	0.988674	0.998738	0.998799
2.3	0.998857	0.998912	0.998966	0.999016	0.999065	0.999111	0.999155	0.999197	0.999237	0.999275
2.4	0.99931	0.999346	0.999379	0.999411	0.999441	0.999469	0.999497	0.999523	0.999547	0.999571
2.5	0.999593	0.999614	0.999635	0.999654	0.999672	0.999689	0.999706	0.999722	0.999736	0.999751
2.6	0.999764	0.999777	0.999789	0.9998	0.999811	0.999822	0.999831	0.999841	0.999849	0.999858
2.7	0.999866	0.999873	0.99988	0.999887	0.999893	0.999899	0.999905	0.99991	0.999916	0.99992
2.8	0.999925	0.999929	0.999933	0.999937	0.999941	0.999944	0.999948	0.999951	0.999954	0.999956
2.9	0.999959	0.999961	0.999964	0.999966	0.999968	0.99997	0.999972	0.999973	0.999975	0.999976
3	0.99997791	0.99997926	0.99998053	0.99998173	0.99998286	0.99998392	0.99998492	0.99998586	0.99998674	0.99998757
3.1	0.99998835	0.99998908	0.99998977	0.99999042	0.99999103	0.9999916	0.99999214	0.99999264	0.99999311	0.99999356
3.2	0.99999397	0.99999436	0.99999473	0.99999507	0.9999954	0.9999957	0.99999598	0.99999624	0.99999649	0.99999672
3.3	0.99999694	0.99999715	0.99999734	0.99999751	0.99999768	0.99999784	0.99999798	0.99999812	0.99999825	0.99999837
3.4	0.99999848	0.99999858	0.99999868	0.99999877	0.99999885	0.99999893	0.99999901	0.99999908	0.99999914	0.9999992
3.5	0.99999926	0.99999931	0.99999936	0.9999994	0.99999945	0.99999948	0.99999952	0.99999956	0.99999959	0.99999962
3.6	0.99999964	0.99999967	0.99999969	0.99999972	0.99999974	0.99999976	0.99999977	0.99999979	0.99999981	0.99999982
3.7	0.99999983	0.99999985	0.99999986	0.99999987	0.99999988	0.99999989	0.99999989	0.9999999	0.99999991	0.99999992
3.8	0.99999992	0.99999993	0.99999993	0.99999994	0.99999994	0.99999995	0.99999995	0.99999996	0.99999996	0.99999996
3.9	0.99999997	0.99999997	0.99999997	0.99999997	0.99999997	0.99999998	0.99999998	0.99999998	0.99999998	0.99999998
4	0.99999998	0.99999999	0.99999999	0.99999999	0.99999999	0.99999999	0.99999999	0.99999999	0.99999999	0.99999999

Index